Springer Tracts in Advanced Robotics

Volume 23

Editors: Bruno Siciliano · Oussama Khatib · Frans Groen

Springer Tracts in Advanced Robotics

Edited by B. Siciliano, O. Khatib, and F. Groen

Vol. 22: Christensen, H.I. (Ed.)
European Robotics Symposium 2006
209 p. 2006 [3-540-32688-X]

Vol. 21: Ang Jr., H.; Khatib, O. (Eds.)
Experimental Robotics IX
618 p. 2006 [3-540-28816-3]

Vol. 20: Xu, Y.; Ou, Y.
Control of Single Wheel Robots
188 p. 2005 [3-540-28184-3]

Vol. 19: Lefebvre, T.; Bruyninckx, H.; De Schutter, J.
Nonlinear Kalman Filtering for Force-Controlled Robot Tasks
280 p. 2005 [3-540-28023-5]

Vol. 18: Barbagli, F.; Prattichizzo, D.; Salisbury, K. (Eds.)
Multi-point Interaction with Real and Virtual Objects
281 p. 2005 [3-540-26036-6]

Vol. 17: Erdmann, M.; Hsu, D.; Overmars, M.; van der Stappen, F.A (Eds.)
Algorithmic Foundations of Robotics VI
472 p. 2005 [3-540-25728-4]

Vol. 16: Cuesta, F.; Ollero, A.
Intelligent Mobile Robot Navigation
224 p. 2005 [3-540-23956-1]

Vol. 15: Dario, P.; Chatila R. (Eds.)
Robotics Research – The Eleventh International Symposium
595 p. 2005 [3-540-23214-1]

Vol. 14: Prassler, E.; Lawitzky, G.; Stopp, A.; Grunwald, G.; Hägele, M.; Dillmann, R.; Iossifidis. I. (Eds.)
Advances in Human-Robot Interaction
414 p. 2005 [3-540-23211-7]

Vol. 13: Chung, W.
Nonholonomic Manipulators
115 p. 2004 [3-540-22108-5]

Vol. 12: Iagnemma K.; Dubowsky, S.
Mobile Robots in Rough Terrain – Estimation, Motion Planning, and Control with Application to Planetary Rovers
123 p. 2004 [3-540-21968-4]

Vol. 11: Kim, J.-H.; Kim, D.-H.; Kim, Y.-J.; Seow, K.-T.
Soccer Robotics
353 p. 2004 [3-540-21859-9]

Vol. 10: Siciliano, B.; De Luca, A.; Melchiorri, C.; Casalino, G. (Eds.)
Advances in Control of Articulated and Mobile Robots
259 p. 2004 [3-540-20783-X]

Vol. 9: Yamane, K.
Simulating and Generating Motions of Human Figures
176 p. 2004 [3-540-20317-6]

Vol. 8: Baeten, J.; De Schutter, J.
Integrated Visual Servoing and Force Control
198 p. 2004 [3-540-40475-9]

Vol. 7: Boissonnat, J.-D.; Burdick, J.; Goldberg, K.; Hutchinson, S. (Eds.)
Algorithmic Foundations of Robotics V
577 p. 2004 [3-540-40476-7]

Vol. 6: Jarvis, R.A.; Zelinsky, A. (Eds.)
Robotics Research – The Tenth International Symposium
580 p. 2003 [3-540-00550-1]

Vol. 5: Siciliano, B.; Dario, P. (Eds.)
Experimental Robotics VIII
685 p. 2003 [3-540-00305-3]

Vol. 4: Bicchi, A.; Christensen, H.I.; Prattichizzo, D. (Eds.)
Control Problems in Robotics
296 p. 2003 [3-540-00251-0]

Vol. 3: Natale, C.
Interaction Control of Robot Manipulators – Six-degrees-of-freedom Tasks
120 p. 2003 [3-540-00159-X]

Vol. 2: Antonelli, G.
Underwater Robots – Motion and Force Control of Vehicle-Manipulator Systems
209 p. 2003 [3-540-00054-2]

Vol. 1: Caccavale, F.; Villani, L. (Eds.)
Fault Diagnosis and Fault Tolerance for Mechatronic Systems – Recent Advances
191 p. 2002 [3-540-44159-X]

Juan Andrade-Cetto · Alberto Sanfeliu

Environment Learning for Indoor Mobile Robots

A Stochastic State Estimation Approach to Simultaneous Localization and Map Building

With 63 Figures

Springer

Professor Bruno Siciliano, Dipartimento di Informatica e Sistemistica, Università degli Studi di Napoli Federico II, Via Claudio 21, 80125 Napoli, Italy, email: siciliano@unina.it

Professor Oussama Khatib, Robotics Laboratory, Department of Computer Science, Stanford University, Stanford, CA 94305-9010, USA, email: khatib@cs.stanford.edu

Professor Frans Groen, Department of Computer Science, Universiteit van Amsterdam, Kruislaan 403, 1098 SJ Amsterdam, The Netherlands, email: groen@science.uva.nl

Authors

Dr. Juan Andrade-Cetto
Prof. Alberto Sanfeliu

Institut de Robòtica i Informàtica Industrial
Universitat Politècnica de Catalunya –
Consejo Superior de Investigaciones Científicas
Llorens Artigas 4–6
08028 Barcelona
Spain

cetto@iri.upc.edu
sanfeliu@iri.upc.edu

ISSN print edition: 1610-7438
ISSN electronic edition: 1610-742X

ISBN 978-3-642-06931-4 e-ISBN 978-3-540-32848-3

Springer is a part of Springer Science+Business Media
springeronline.com

Printed in Germany

Cover-Design: design & production GmbH, Heidelberg
Printed on acid-free paper 89/3141/Yu - 5 4 3 2 1 0

STAR (Springer Tracts in Advanced Robotics) has been promoted under the auspices of EURON (European Robotics Research Network)

To Irene and María, *JAC*
To Ana, Alberto, Cristina, and Elena, *AS*

Foreword

At the dawn of the new millennium, robotics is undergoing a major transformation in scope and dimension. From a largely dominant industrial focus, robotics is rapidly expanding into the challenges of unstructured environments. Interacting with, assisting, serving, and exploring with humans, the emerging robots will increasingly touch people and their lives.

The goal of the new series of Springer Tracts in Advanced Robotics (STAR) is to bring, in a timely fashion, the latest advances and developments in robotics on the basis of their significance and quality. It is our hope that the wider dissemination of research developments will stimulate more exchanges and collaborations among the research community and contribute to further advancement of this rapidly growing field.

The monograph written by Juan Andrade-Cetto and Alberto Sanfeliu is focused on a popular research topic in the latest few years, namely Simultaneous Localization and Map Bulding (SLAM). The estimation theoretical aspects are covered with resort to the widely-adopted Extended Kalman Filtering (EKF) technique. Further to the design of the estimator, the controller design is also discussed in the work along with its implications on closing the perception-action loop. Both simulation and experimental results for indoor mobile robots are presented to show the effectiveness of the proposed methods.

Remarkably, the doctoral thesis at the basis of this monograph received the prize of the Fourth Edition of the EURON Georges Giralt PhD Award devoted to the best PhD thesis in Robotics in Europe. A fine addition to the Series!

Naples, Italy
October 2005

Bruno Siciliano
STAR Editor

Preface

Efficient mobile robot navigation is limited mainly by the ability of a robot to perceive and interact with its surroundings in a deliberative way. A desirable characteristic a mobile robot must have are the skills needed to recognize the landmarks and objects that surround it, and to be able to localize itself relative to its workspace. This knowledge is crucial for the successful completion of intelligent navigation tasks. But, for such interaction to take place, a model or description of the environment needs to be specified beforehand.

If a global description or measurement of the elements present in the environment is available, the problem consists on the interpretation and matching of sensor readings to such previously stored object models. Moreover, if we know that the recognized objects are fixed and persist in the scene, they can be regarded as landmarks, and can be used as reference points for self-localization. If on the other hand, a global description or measurement of the elements in the environment is not available, at least the descriptors and methods that will be used for the autonomous building of one are required. This is, either the robot has a global map, or it is given the means to learn one.

We are interested in this second case. That is, in providing an autonomous robot with the necessary skills to build a map and precisely localize itself within this map while navigating in previously unexplored settings. The research reported in this monograph focuses on some estimation theoretic aspects of the so called Simultaneous Localization and Map Building (SLAM) problem.

We start our discussion by reviewing in Chapter 1 the traditional full covariance extended Kalman filter approach to simultaneous localization and map building (EKF-SLAM). Explicit formulas for two mobile platforms are presented. First, we show the case of a simple linear one-dimensional mobile robot, the *monobot*. Then, we extend the analysis to the more realistic case of a planar mobile robot.

At the end of Chapter 1 we introduce a pair of temporal landmark quality functions to aid in those situations in which landmark observations might not be consistent in time; and show how by incorporating these functions, the overall estimation-theoretic approach to SLAM is improved. Special attention is paid in that the removal of landmarks from the map does not violate the basic convergence properties of the localization and map building algorithms already described in the literature. Namely, asymptotic convergence and full correlation.

Chapter 2 presents an analysis of the fully correlated approach to SLAM from a control systems theory point of view, both for linear and nonlinear vehicle models. It shows how partial observability hinders full reconstructibility of the state space, making the final map estimate dependent on the initial observations. Nevertheless, marginal filter stability guarantees convergence of the state error covariance to a positive semi-definite covariance matrix. By characterizing the form of the total Fisher information we are able to determine the unobservable state space directions. Moreover, we give a closed form expression that links the amount of reconstruction error to the number of landmarks used. The analysis allows the formulation of measurement models that make SLAM observable.

In the search for real-time implementations of SLAM, covariance inflation methods produce a suboptimal filter that eventually may lead to the computation of an unbounded state error covariance. Chapter 3 provides tight constraints in the amount of decorrelation possible, to guarantee convergence of the state error covariance, and at the same time, a linear-time implementation of SLAM.

In Chapter 4 we propose an algorithm to reduce the effects caused by linearization in the typical EKF approach to SLAM. The technique consists in computing the vehicle prior using an Unscented Transformation. The UT allows a better nonlinear mean and variance estimation than the EKF. There is no need however in using the UT for the entire vehicle-map state, given the linearity in the map part of the model. By applying the UT only to the vehicle states we get more accurate covariance estimates. The a posteriori estimation is made using a fully observable EKF step, thus preserving the same computational complexity as the EKF with sequential innovation. Experiments over a standard SLAM data set show the behavior of the algorithm.

The last chapter is about closing the low level control loop during Simultaneous Localization and Map Building from an estimation-control theoretic viewpoint. We present first, the case of an optimal state regulator for the linear SLAM case, commonly referred as Linear Quadratic Regulator, and show also its behavior in the case of the EKF. Then we present a feedback linearization multi-vehicle control strategy that uses the state estimates generated from the SLAM algorithm as input to a multi-vehicle controller. Given the separability between optimal state estimation and regulation, we show that the tracking error does not influence the estimation performance of a fully observable EKF based multirobot SLAM implementation, and viceversa, that estimation errors do not undermine controller performance. Furthermore, both the controller and estimator are shown to be asymptotically stable. The feasibility of using this technique to close the perception-action loop during multirobot SLAM is validated with simulation results.

The first two chapters of this monograph derive from the PhD work of the first author. The last three chapters come from the continuation of his research endeavors first, while in a postdoctoral stay at IRI, and later on, as a Juan de la Cierva posdoctoral fellow at CVC, UAB. Many of the ideas in those chapters are due also to Teresa Vidal-Calleja, currently a PhD student at IRI under our supervision. This monograph is being published in the Springer Tracts in Advanced Robotics Series upon reception of the EURON Georges Giralt Best PhD Award in 2005.

The work reported here was supported in part by the Spanish Ministry of Education and Science under projects DPI 2001-2223, TIC 2003-09291 and DPI 2004-5414.

Barcelona, Spain
October 2005

Juan Andrade-Cetto
Alberto Sanfeliu

Contents

1

Simultaneous Localization and Map Building

Perception happens locally, in the egocentric frame of reference of the robot. In order to ensure correspondence between the local representation of the environment built by the landmark extraction processes, and the global representation contained in a map, the robot must estimate its own position with respect to this map.

The use of stochastic models for map building and localization in mobile robotics has gained much popularity in recent years [24, 38, 61]. Of particular interest is the use of predictive filters to estimate the robot position and uncertainty, and to update these estimates from sensor readings while at the same time building an incremental map of the environment [7, 16, 21, 31, 63, 79, 83].

One of the most critical limitations to the application of such estimation-theoretic approaches to map building and localization is the data association problem. Data association refers to the issue of matching observations with previously learned elements from the environment. Some techniques can be used to alleviate the data association problem, such as the tracking of landmarks from one robot position to the next, or by using efficient tests for scene-to-model landmark match hypothesis verification. Obviously there is always a compromise between the possibility of fully invariant landmark characterization and the difficulty to extract such characterizing features from raw sensor data.

As we address issues such as viewpoint invariance and feature extraction from sensor data, it is overwhelming how undesired environment dynamics, occlusions, and sensor noise can still make data association a daunting task. One possibility to overcome the data association problem altogether is with the deployment of uniquely identifiable man-made beacons to aid in localization. Unfortunately, there exist multiple situations where this is not possible, and a map must still be constructed without environment contamination. An alternative approach explored in this work is the use of temporal and spatial landmark quality measures to validate observations.

During the course of our research we have tested and implemented SLAM solutions for indoor mobile robots with a laser range scanner, based mainly on the algorithm described in this chapter, and with the extensions described in subsequent chapters along this book, such as full observability, temporal landmark quality, unscented vehicle transformations, etc. Figure 1.1 plots a series of snapshots of a test run of our SLAM algorithm with our mobile robot Marco from Figure 1.2. A 3d partial representation of the final map built is shown in Figure 1.3. In the plots the reader can see for example, localization variances as level curves around wall endpoints. These are the type of maps one could expect to obtain when using an EKF approach to SLAM, such as the one explained throgout this book.

We start our discussion by reviewing in Section 1.1 the traditional full covariance extended Kalman filter approach to simultaneous localization and map building (EKF-SLAM in short), based primarily on the works by Smith and Cheeseman [79] and Dissanayake *et al.* [31]. In Section 1.2, explicit formulas for two mobile platforms are presented. First, we show the case of a simple linear one-dimensional mobile robot, the *monobot.* Then, we extend the analysis to the more realistic case of a planar mobile robot.

Spatial landmark compatibility tests are needed to validate data association hypothesis in terms of the estimated localization error for each landmark. Their use is crucial for the solution of data association in SLAM [21, 70]. We have realized however, that in situations with moderate scene dynamics, spatial landmark compatibility may not suffice in the search for data association matches. Consider for example the case when a landmark is occluded for a short period of time. A spatial compatibility test would not have any information on the history of observations of such landmark, and might still be trying to wrongly associate it with a neighboring observed feature. If the algorithm succeeds in incorrectly associating the occluded feature, the new observation will not be consistent with the initial measurement, thus producing large error in the estimate for the localization of that landmark, while at the same time underestimating its covariance. Given that the map covariance is fully correlated, starting with the next iteration of the algorithm, that wrong value for the uncertainty would be propagated to the rest of the landmark locations, and that of the robot as well; leading to divergence in the map, and ultimately breaking down the entire estimation approach to SLAM.

To aid in those situations in which landmark observations might not be consistent in time, we propose a new set of temporal landmark quality models, and show how by incorporating these models, the overall estimation-theoretic approach to SLAM is improved. With the aid of these models, a new temporal landmark quality test is presented to aid in differentiating between the imprecision in the localization of a landmark, and the uncertainty in its very existence. Thanks to this test we are able to remove weak landmarks from the map. Landmarks that would most likely be a product of false data association or spurious observations, and that if considered, would otherwise induce

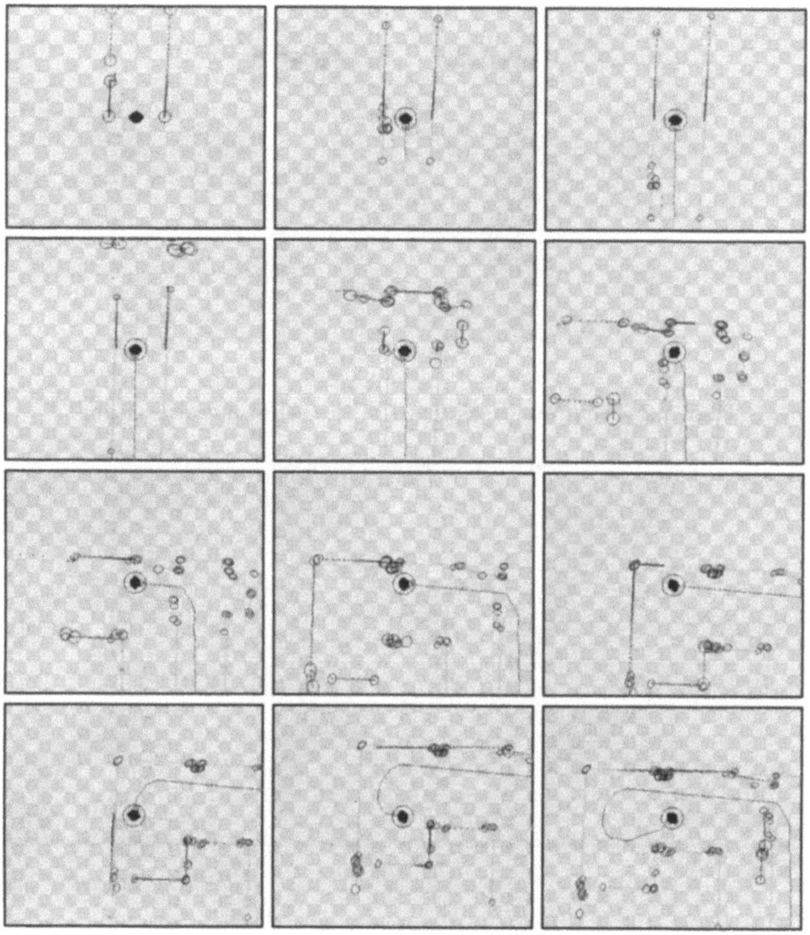

Fig. 1.1. The blue dots indicate sensor raw data coming from a laser range finder, and the blue ellipses represent 2σ confidence level curves on the wall end point estimates. The green lines represent walls inferred from consecutive readings. The red lines indicate the estimated robot trajectory.

undesired localization errors. Temporal landmark compatibility is addressed in Section 1.3.

Finally, in Section 1.4, our planar mobile robot configuration is used to evaluate the original full-covariance Extended Kalman Filter algorithm to Simultaneous Localization and Map Building as reported by Dissanayake *et al* [31], including the spatial landmark compatibility tests [70], versus our improved algorithm, the EKF-SLAM-LV, with both temporal and spatial landmark quality tests, both in the presence of various noise levels, and ultimately, in cases with limited field of view and extreme data missassociation.

Fig. 1.2. Front and back views of the robot initial position.

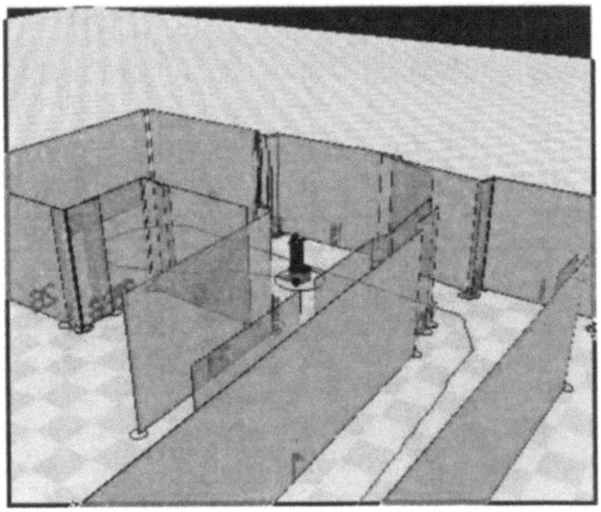

Fig. 1.3. Graphical representation of the map built.

1.1 Extended Kalman Filter Approach to SLAM

In this Section we review the fundamentals of the stochastic estimation approach to simultaneous localization and map building. The material covered summarizes the work of many researchers during the past 15 years, and will constitute a starting point for our view of the mobile robot localization and map building problem.

Full covariance EKF SLAM

Before delving into the mathematical formulation that builds up the full covariance Extended Kalman Filter approach to Simultaneous Localization and

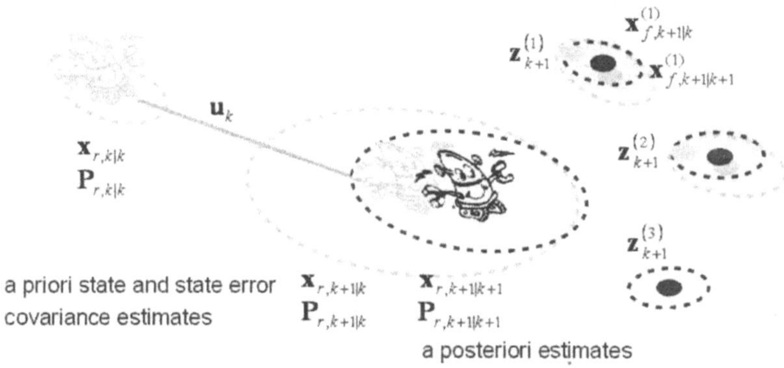

Fig. 1.4. State estimation approach to simultaneous localization and map building.

Map Building, we proceed with a motivational explanation of how such a predictive filter can be used to solve the localization problem.

Imagine, for the sake of simplicity, a planar mobile robot that we believe is located at position $\mathbf{x}_{r,k|k}$ as shown in Figure 1.4; and that such location is known with some level of uncertainty indicated by the area inside the ellipse $\mathbf{P}_{r,k|k}$. The subscript r indicates the part of the state vector $\mathbf{x}$ and of the error covariance matrix $\mathbf{P}$ associated to the robot pose. The ellipse indicates a level curve of equal uncertainty, its principal axes are the eigenvectors of the covariance matrix $\mathbf{P}_r$. The term k is used to indicate the time stamp.

Driving the robot with the motion command $\mathbf{u}_k$, we would expect it to end up at location $\mathbf{x}_{r,k+1|k}$. Such location estimate, the *a priori* state estimate, posses a larger level of uncertainty than our previous estimate $\mathbf{x}_{r,k|k}$. The reason, is the addition of dead reckoning error and the inaccuracies of our motion model. This increase in uncertainty is exemplified by the larger area inside the ellipse $\mathbf{P}_{r,k+1|k}$.

Once the motion command is complete, the robot sensors acquire information about the environment. More specifically, they measure the distance from the robot to a number of fixed landmarks $\mathbf{x}_f^{(1)}$, $\mathbf{x}_f^{(2)}$, and $\mathbf{x}_f^{(3)}$. It is from these observations, that the filter corrects its estimate about the location of the robot to $\mathbf{x}_{r,k+1|k+1}$, the *a posteriori* estimate; and at the same time, reduces the localization uncertainty $\mathbf{P}_{r,k+1|k+1}$. The true robot location $\mathbf{x}_{r,k+1}$ is still unknown; however, the idea behind the use of a predictive filter is to minimize the estimation error $(\mathbf{x}_{r,k+1} - \mathbf{x}_{r,k+1|k+1})$. In other words, we want an estimator that keeps the uncertainty ellipse for $\mathbf{P}_{r,k+1|k+1}$ as small as possible.

Fortunately, we can resort to the Kalman filter, a recursive stochastic state estimator for partially observed non-stationary processes that gives an optimal state estimate in the least squares sense. In the typical full-covariance EKF based approach to SLAM we use such a filter precisely to iteratively estimate the robot and landmark locations, optimally reducing the error between the true robot location and our computed estimate.

System model

Formally speaking, the motion of the robot and the measurement of the map features are governed by the discrete-time state transition model

$$\mathbf{x}_{k+1} = \mathbf{f}(\mathbf{x}_k, \mathbf{u}_k, \mathbf{v}_k) \tag{1.1}$$

$$\mathbf{z}_k = \mathbf{h}(\mathbf{x}_k) + \mathbf{w}_k \tag{1.2}$$

The state vector $\mathbf{x}_k \in \mathbb{R}^{m+dn}$ contains the position of the robot $\mathbf{x}_{r,k} \in \mathbb{R}^m$ at time step k, and a vector of n stationary d-dimensional map features $\mathbf{x}_f \in \mathbb{R}^{dn}$, i.e.,

$$\mathbf{x}_k = \begin{bmatrix} \mathbf{x}_{r,k} \\ \mathbf{x}_f \end{bmatrix} \tag{1.3}$$

The input vector $\mathbf{u}_k \in \mathbb{R}^m$ is the vehicle control command, and $\mathbf{v}_k \in \mathbb{R}^m$ is a Gaussian random vector with zero mean and covariance matrix $\mathbf{Q}_k \in \mathbb{R}^{m\times m}$, representing unmodeled robot dynamics and system noise. The function $\mathbf{f} : \mathbb{R}^{m+dn} \to \mathbb{R}^{m+dn}$ is a possibly nonlinear difference equation that models the motion of the robot.

The Gaussian random vector $\mathbf{w}_k \in \mathbb{R}^{dn}$ represents both, the inaccuracies of the also possibly nonlinear observation model $\mathbf{h} : \mathbb{R}^{dn+m} \to \mathbb{R}^{dn}$, and the measurement noise with zero mean and covariance matrix $\mathbf{R}_k \in \mathbb{R}^{dn\times dn}$.

Provided the set of observations $Z^k = \{\mathbf{z}_1, \ldots, \mathbf{z}_k\}$ was available for the computation of the current map estimate $\mathbf{x}_{k|k}$, the expression

$$\mathbf{x}_{k+1|k} = \mathbf{f}(\mathbf{x}_{k|k}, \mathbf{u}_k, \mathbf{0}) \tag{1.4}$$

gives an a priori noise-free estimate of the new locations of the robot and map features after the vehicle control command $\mathbf{u}_k$ is input to the system. Similarly,

$$\mathbf{z}_{k+1|k} = \mathbf{h}(\mathbf{x}_{k+1|k}, \mathbf{0}) \tag{1.5}$$

constitutes a noise-free a priori estimate of sensor measurements.

The EKF approach to SLAM requires the linearization of both the motion and observation models. Such linearizations are formulated as Taylor series approximations with the higher order terms dropped, i.e.,

$$\mathbf{x}_{k+1} \approx \mathbf{x}_{k+1|k} + \mathbf{F}(\mathbf{x}_k - \mathbf{x}_{k|k}) + \mathbf{G}\mathbf{v}_k \tag{1.6}$$

$$\mathbf{z}_{k+1} \approx \mathbf{z}_{k+1|k} + \mathbf{H}(\mathbf{x}_{k+1} - \mathbf{x}_{k+1|k}) + \mathbf{w}_{k+1} \tag{1.7}$$

The Jacobian matrices $\mathbf{F}$, $\mathbf{G}$, and $\mathbf{H}$ contain the partial derivatives of $\mathbf{f}$ with respect to $\mathbf{x}$ and the noise $\mathbf{v}$, and of $\mathbf{h}$ with respect to $\mathbf{x}$, respectively:

$$\mathbf{F} = \left.\frac{\partial \mathbf{f}}{\partial \mathbf{x}}\right|_{(\mathbf{x}_{k|k},\mathbf{u}_k,\mathbf{0})} \tag{1.8}$$

$$\mathbf{G} = \left.\frac{\partial \mathbf{f}}{\partial \mathbf{v}}\right|_{(\mathbf{x}_{k|k},\mathbf{u}_k,\mathbf{0})} \tag{1.9}$$

$$\mathbf{H} = \left.\frac{\partial \mathbf{h}}{\partial \mathbf{x}}\right|_{(\mathbf{x}_{k+1|k},\mathbf{0})} \tag{1.10}$$

Given that the landmarks are considered stationary, their a priori estimate is simply

$$\mathbf{x}_{f,k+1|k} = \mathbf{x}_{f,k|k} \tag{1.11}$$

Thus, the elements of the non-stationary linear state transition model of the robot and map dynamics in Equations 1.6 and 1.7 take the forms

$$\begin{bmatrix} \mathbf{x}_{r,k+1} \\ \mathbf{x}_f \end{bmatrix} \approx \begin{bmatrix} \mathbf{x}_{r,k+1|k} \\ \mathbf{x}_{f,k|k} \end{bmatrix} + \begin{bmatrix} \mathbf{F}_r & \\ & \mathbf{I} \end{bmatrix} \begin{bmatrix} \tilde{\mathbf{x}}_{r,k|k} \\ \tilde{\mathbf{x}}_{f,k|k} \end{bmatrix} + \begin{bmatrix} \mathbf{G}_r & \\ & \mathbf{0} \end{bmatrix} \begin{bmatrix} \mathbf{v}_k \\ \mathbf{0} \end{bmatrix} \tag{1.12}$$

$$\mathbf{z}_{k+1} \approx \mathbf{z}_{k+1|k} + \begin{bmatrix} \mathbf{H}_r & \mathbf{H}_f \end{bmatrix} \begin{bmatrix} \tilde{\mathbf{x}}_{r,k+1|k} \\ \tilde{\mathbf{x}}_{f,k+1|k} \end{bmatrix} + \mathbf{w}_{k+1} \tag{1.13}$$

with the *a priori state error*

$$\begin{aligned} \tilde{\mathbf{x}}_{k+1|k} &= \begin{bmatrix} \tilde{\mathbf{x}}_{r,k+1|k} \\ \tilde{\mathbf{x}}_{f,k+1|k} \end{bmatrix} \\ &= \begin{bmatrix} \mathbf{x}_{r,k+1} - \mathbf{x}_{r,k+1|k} \\ \mathbf{x}_f - \mathbf{x}_{f,k+1|k} \end{bmatrix} \end{aligned} \tag{1.14}$$

and the *a posteriori state error*

$$\begin{aligned} \tilde{\mathbf{x}}_{k|k} &= \begin{bmatrix} \tilde{\mathbf{x}}_{r,k|k} \\ \tilde{\mathbf{x}}_{f,k|k} \end{bmatrix} \\ &= \begin{bmatrix} \mathbf{x}_{r,k} - \mathbf{x}_{r,k|k} \\ \mathbf{x}_f - \mathbf{x}_{f,k|k} \end{bmatrix} \end{aligned} \tag{1.15}$$

In the remaining of this chapter, extensive use of Kalman Filter related notation will be used. A few terms the reader might want to keep in mind include: the error covariance matrix $\mathbf{P}$, the measurement innovation covariance matrix $\mathbf{S}$, and the Kalman gain $\mathbf{K}$. If in doubt, please refer to Appendix A.

Algorithm

An a priori prediction of the location of the robot and the state of the map is computed in Equation 1.4 purely from motion commands; consequently increasing the uncertainty of the robot location and that of the map features.

In general terms, the a priori estimate to the vehicle and map state error covariance showing this increase in uncertainty is given by

$$\begin{aligned}\mathbf{P}_{k+1|k} &= E[\tilde{\mathbf{x}}_{k+1|k}\,\tilde{\mathbf{x}}_{k+1|k}^\top] \\ &= \mathbf{F}\mathbf{P}_{k|k}\mathbf{F}^\top + \mathbf{G}\mathbf{Q}_k\mathbf{G}^\top \end{aligned} \tag{1.16}$$

Writing the map state error covariance matrix in block form, and substituting the corresponding Jacobian matrices, we can rewrite Equation 1.16 as

$$\begin{bmatrix} \mathbf{P}_{r,k+1|k} & \mathbf{P}_{rf,k+1|k} \\ \mathbf{P}^\top_{rf,k+1|k} & \mathbf{P}_{f,k+1|k} \end{bmatrix} = \begin{bmatrix} \mathbf{F}_r\mathbf{P}_{r,k|k}\mathbf{F}_r^\top + \mathbf{G}_r\mathbf{Q}_k\mathbf{G}_r{}^\top & \mathbf{F}_r\mathbf{P}_{rf,k|k} \\ \mathbf{P}^\top_{rf,k|k}\mathbf{F}_r^\top & \mathbf{P}_{f,k|k} \end{bmatrix} \tag{1.17}$$

Assuming that a new set of landmark observations $\mathbf{z}_{k+1}$ coming from sensor data has been correctly matched to their map counterparts, one can compute the error between the measurements and the estimates with

$$\tilde{\mathbf{z}}_{k+1|k} = \mathbf{z}_{k+1} - \mathbf{z}_{k+1|k} \tag{1.18}$$

This error aids in revising the map and robot locations. The a posteriori state estimate is

$$\mathbf{x}_{k+1|k+1} = \mathbf{x}_{k+1|k} + \mathbf{K}_{k+1}\tilde{\mathbf{z}}_{k+1|k} \tag{1.19}$$

and the Kalman gain is computed with

$$\mathbf{K}_{k+1} = \mathbf{P}_{k+1|k}\mathbf{H}^\top\mathbf{S}^{-1} \tag{1.20}$$

where $\mathbf{S}$ is the measurement innovation matrix,

$$\mathbf{S} = \mathbf{H}\mathbf{P}_{k+1|k}\mathbf{H}^\top + \mathbf{R}_{k+1} \tag{1.21}$$

Finally, the a posteriori estimate of the map state error covariance must also be revised once a measurement has taken place. It is revised with

$$\mathbf{P}_{k+1|k+1} = (\mathbf{I} - \mathbf{K}_{k+1}\mathbf{H})\,\mathbf{P}_{k+1|k} \tag{1.22}$$

or equivalently, and to guarantee positive semi-definiteness of $\mathbf{P}_{k+1|k+1}$, with

$$\mathbf{P}_{k+1|k+1} = (\mathbf{I} - \mathbf{K}_{k+1}\mathbf{H})\,\mathbf{P}_{k+1|k}\,(\mathbf{I} - \mathbf{K}_{k+1}\mathbf{H})^\top + \mathbf{K}_{k+1}\mathbf{R}_{k+1}\mathbf{K}_{k+1}^\top \tag{1.23}$$

The above expression is commonly referred as the Joseph form of the a posteriori state error covariance matrix. Its derivation is discussed in detail in Appendix A. The properties of positive semi-definite (psd) matrices are enumerated in Section B for completeness of the discussion.

The contribution to the revision of the robot pose and landmark location estimates is proportional to our degree of trust in the motion and sensor models respectively. If the plant error covariance $\mathbf{Q}$ is large, and the measurement error covariance $\mathbf{R}$ is small, the EKF-SLAM algorithm trusts more the observations than dead-reckoning, revising more heavily the robot pose estimate than that of the landmarks. Conversely, when the measurement error covariance is larger than the plant error covariance, the algorithm trusts more on the motion of the robot and ends up revising more heavily the landmark estimates.

Convergence properties

One important property of the estimation-theoretic approach to SLAM is that the map is asymptotically convergent. That is, in the original full covariance KF-based SLAM formulation the map state error covariance submatrix associated with the landmark estimates decreases monotonically as successive observations take place. Formally speaking,

$$\det \mathbf{P}_{f,k+1|k+1} \leq \det \mathbf{P}_{f,k|k} \tag{1.24}$$

Another property indicates how *in the limit*, as the number of iterations tends to infinity, the map becomes fully correlated; suggesting that if a landmark location is given, the location of the other landmarks can be deduced with absolute certainty from the map built.

$$\lim_{k\to\infty} \det \mathbf{P}_{f,k|k} = 0 \tag{1.25}$$

The third property in the analysis of the covariance in EKF-SLAM, is that in the limit, the absolute location of the vehicle and map is bounded by the initial vehicle uncertainty. And for a vehicle with no process noise

$$\lim_{k\to\infty} \mathbf{P}_{r,k|k} = \mathbf{P}_{r,0|0} \tag{1.26}$$

These properties were first reported in Newman's PhD work [31, 71], and will be of use in the following Chapter where we talk about partial observability, partial controllability, and filter stability in SLAM.

Sequential innovation

EKF-SLAM requires that all landmarks in the map be always observed by the vehicle, and be correctly associated to their map counterpart at every iteration. However, it is also possible to revise the state estimate only with partial observations. Independent landmark measurements contribute only to the revision of the map states directly associated to that particular landmark; and, if a sufficient number of independent landmark observations are made, it is still possible to have the system reliably estimate the location of the robot and landmarks iteratively. The technique used to update the state estimate one observation at a time is called *sequential innovation* (see Appendix A). The result of applying sequential innovation to all the landmarks in the map is equivalent to that of using the full covariance extended Kalman filter approach, provided the observations are independent, or that at least they can be *whitened* [93].

The main advantage of using sequential innovation, is that by considering the measurement vector $\mathbf{z}_{k+1}$ as a set of single measurements $\mathbf{z}^{(1)}_{k+1} \dots \mathbf{z}^{(n)}_{k+1}$ that can be treated sequentially, the inversion of the joint measurement innovation covariance matrix $\mathbf{S}$ is no longer necessary. Instead, a series of smaller individual innovation covariance matrix inverses is computed, reducing considerably the time complexity of the algorithm.

Starting from $\mathbf{x}^{(0)}_{k+1,k+1} = \mathbf{x}_{l+1|k}$, and $\mathbf{P}^{(0)}_{k+1|k+1} = \mathbf{P}_{k+1|k}$, the a posteriori state estimate is iteratively given by

$$\mathbf{x}^{(i)}_{k+1|k+1} = \mathbf{x}^{(i-1)}_{k+1|k+1} + \mathbf{K}_{(i)}(\tilde{\mathbf{z}}^{(i)}_{k+1|k}). \tag{1.27}$$

with $\mathbf{K}_{(i)}$ the i-th set of columns of $\mathbf{K}$.

Individual landmark measurements can be estimated taking only the corresponding i-th set of rows of the measurement Jacobian $\mathbf{H}$

$$\mathbf{z}^{(i)}_{k+1|k} = \mathbf{H}^{(i)}\mathbf{x}^{(i-1)}_{k+1|k+1} \tag{1.28}$$

whit individual landmark estimation errors

$$\tilde{\mathbf{z}}^{(i)}_{k+1|k} = \mathbf{z}^{(i)}_{k+1} - \mathbf{z}^{(i)}_{k+1|k} \tag{1.29}$$

The independent treatment of the observations is only possible if $\mathbf{R}$ is block diagonal. This is, when the landmark observations taken during the same time interval are uncorrelated. The i-th set of columns of $\mathbf{K}$ is simply computed with

$$\mathbf{K}_{(i)} = \mathbf{P}^{(i-1)}_{k+1|k+1}\mathbf{H}^{(i)^\top}\mathbf{S}^{(i)^{-1}} \tag{1.30}$$

with the *individual innovation covariance matrices*

$$\mathbf{S}^{(i)} = \mathbf{H}^{(i)}\mathbf{P}^{(i-1)}_{k+1|k+1}\mathbf{H}^{(i)^\top} + \mathbf{R}^{(i)} \tag{1.31}$$

The a posteriori state error covariance for individual observations is also iteratively computed with

$$\mathbf{P}^{(i)}_{k+1|k+1} = \mathbf{P}^{(i-1)}_{k+1|k+1} - \mathbf{K}_{(i)}\mathbf{H}^{(i)}\mathbf{P}^{(i-1)}_{k+1|k+1} \tag{1.32}$$

The processing of observations for time step $k+1$ ends with $\mathbf{x}_{k+1|k+1} = \mathbf{x}^{(n)}_{k+1|k+1}$ and $\mathbf{P}_{k+1|k+1} = \mathbf{P}^{(n)}_{k+1|k+1}$.

The required inverse in Equation 1.30 is of size d, and is considerably much smaller than the dimensions of the entire measurement vector $\mathbf{z}$, dn, as required in Equation 1.20. However, if one still wishes to compute the inverse of the whole innovation covariance matrix $\mathbf{S}$ for more than one landmark at a time, one can resort to an incremental computation of $\mathbf{S}^{-1}$ as shown in Section 1.1.

Another consequence of using sequential innovation is that the model Jacobians can also be re-computed after each observation is incorporated to the filter, thus producing even more accurate state and state covariance estimates.

Covariance initialization

Another crucial implementation aspect of the full covariance EKF approach to SLAM is the initialization of the error covariance matrix $\mathbf{P}$ as new observations are added to the map. This matrix contains the expected robot and landmark localization error, and will only manifest the asymptotic convergence properties shown in Section 1.1 if it is initialized properly.

The function that maps an observation into world coordinates is given by our linearized measurement model, and has the form

$$\begin{bmatrix} \mathbf{x}_r \\ \mathbf{x}_f^{(i)} \end{bmatrix} = \mathbf{M} \begin{bmatrix} \mathbf{x}_r \\ \mathbf{z}^{(i)} \end{bmatrix} \tag{1.33}$$

$\mathbf{M}$ is known as the *feature initialization* matrix. Solving for $\mathbf{M}$ from Equations 1.5 and 1.7, we get

$$\mathbf{M} = \begin{bmatrix} \mathbf{I} & \mathbf{0} \\ -\mathbf{H}_{f,(i)}^{-1}\mathbf{H}_{r,(i)} & \mathbf{H}_{f,(i)}^{-1} \end{bmatrix} \tag{1.34}$$

Consequently, the initialization of the corresponding map state error covariance for such landmark is given by

$$\mathbf{P} = \mathbf{M} \begin{bmatrix} \mathbf{P}_r & \mathbf{0} \\ \mathbf{0} & \mathbf{Q}^{(i)} \end{bmatrix} \mathbf{M}^\top \tag{1.35}$$

Without loss of generality, assume a map with $n-1$ landmarks at time step k. The map state error covariance matrix has the form

$$\mathbf{P}_{k|k} = \begin{bmatrix} \mathbf{P}_{r,k|k} & \mathbf{P}_{rf^{(1)},k|k} & \cdots & \mathbf{P}_{rf^{(n-1)},k|k} \\ \mathbf{P}^\top_{rf^{(1)},k|k} & \mathbf{P}_{f^{(1)}f^{(1)},k|k} & \cdots & \mathbf{P}_{f^{(1)}f^{(n-1)},k|k} \\ \vdots & & & \\ \mathbf{P}^\top_{rf^{(n-1)},k|k} & \mathbf{P}^\top_{f^{(1)}f^{(n-1)},k|k} & \cdots & \mathbf{P}_{f^{(n-1)}f^{(n-1)},k|k} \end{bmatrix} \tag{1.36}$$

Once the robot has observed a sufficiently robust new feature which cannot be associated to any other landmark in the map, it is labeled as the n-th landmark, and a new row and column must be appended to the map covariance matrix with

$$\mathbf{P}_{rf^{(n)},k|k} = -\mathbf{P}_{r,k|k}(\mathbf{H}_{f,(n)}^{-1}\mathbf{H}_{r,(n)})^\top \tag{1.37}$$

$$\mathbf{P}_{f^{(i)}f^{(n)},k|k} = \mathbf{H}_{f,(i)}^{-1}\mathbf{H}_{r,(i)}\mathbf{P}_{r,k|k}(\mathbf{H}_{f,(n)}^{-1}\mathbf{H}_{r,(n)})^\top \tag{1.38}$$

$$\mathbf{P}_{f^{(n)}f^{(n)},k|k} = \mathbf{H}_{\mathbf{x}_f,(n)}^{-1}\mathbf{H}_{r,(n)}\mathbf{P}_{r,k|k}(\mathbf{H}_{f,(n)}^{-1}\mathbf{H}_{r,(n)})^\top + \mathbf{H}_{f,(n)}^{-1}\mathbf{Q}^{(i)}(\mathbf{H}_{f,(n)}^{-1})^\top \tag{1.39}$$

Equations 1.37-1.39 indicate that the initialization of the new feature map error covariance is a function of the actual vehicle position and its accumulated uncertainty.

Landmark spatial uncertainty

The estimated uncertainty in the localization of every landmark in the map, as well as that of the robot, is maintained in the state error covariance $\mathbf{P}$. Consequently, the uncertainty of its location in observation space is given by the change of basis of $\mathbf{P}$ plus that of the independent measurement uncertainties $\mathbf{R}^{(i)}$. This quantity is called the innovation covariance matrix, and is precisely equivalent to the term already introduced in Equation 1.21.

For any particular measurement $\mathbf{z}^{(i)}_{k+1}$, the squared Mahalanobis distance

$$d_i^2 = \tilde{\mathbf{z}}^{(i)}_{k+1|k}{}^{\top} \mathbf{S}^{(i)}{}^{-1} \tilde{\mathbf{z}}^{(i)}_{k+1|k} \tag{1.40}$$

represents a measure of spatial disparity between the observation $\mathbf{z}^{(i)}_{k+1}$ and the estimated location in robot centered coordinates of the hypothetical landmark match $\mathbf{x}^{(i)}_f$.

Two spatial landmark compatibility tests that appear in the literature are the individual compatibility test and the joint compatibility test. The former considers landmark observations independently and has been widely used not only to validate observations within the framework of the SLAM problem, but on a wide range of applications. A related example in the computer vision literature, is for example, the evaluation of visual correspondences in the computation of the fundamental matrix of the two-view geometry [46]. The joint compatibility test on the other hand, due to Neira and Tardós [70], considers cross correlated landmark uncertainties when testing match hypotheses, at the expense of higher computational cost.

Individual compatibility test

The squared Mahalanobis distance in Equation 1.40 is a weighted squared sum of d Gaussian variables and as such, follows a χ^2 distribution. Moreover, the subset of map features compatible with measurement $\mathbf{z}^{(i)}_{k+1}$ are the ones that satisfy the χ^2 compatibility test

$$d_i^2 \leq \chi^2_{d,\alpha} \tag{1.41}$$

The number of degrees of freedom in the individual χ^2 compatibility test is given by the rank of the one landmark innovation covariance matrix $\mathbf{S}^{(i)}$. Given that $\mathbf{S}^{(i)}$ is invertible, it must have full rank. Consequently, the number of degrees of freedom of the test is d. So for example, if $\mathbf{z}^{(i)}$ is an image point, the squared Mahalanobis distance represents the sum of the squared horizontal and vertical pixel measurement errors, weighted by the covariance matrix $\mathbf{S}^{(i)}$; and in this case, the number of degrees of freedom of the individual compatibility test is 2. The term α indicates the desired confidence level; so for example, a value of $\alpha = 0.95$ indicates a 95% probability that the observation $\mathbf{z}^{(i)}$ matches the map feature $\mathbf{x}^{(i)}_f$. In other words, a 5% probability that correct data associations will not pass the compatibility test. By increasing the confidence level α, we are augmenting the rate of observations that pass the compatibility test, even if they correspond to incorrect matches.

One clear drawback of the individual compatibility test is the fact that no correlational information is used. An expansion of $\mathbf{S}^{(i)}$, reveals how the square Mahalanobis distance d_i^2 is indeed an individual measure of landmark compatibility, in which no landmark correlation information is taken into account.

$$\mathbf{S}^{(i)} = \mathbf{H}_r^{(i)} \mathbf{P}_{r,k+1|k} \mathbf{H}_r^{(i)^\top} + \mathbf{H}_f^{(i)} \mathbf{P}_{rf,k+1|k}^{(i)}{}^\top \mathbf{H}_r^{(i)^\top} + \mathbf{H}_r^{(i)} \mathbf{P}_{rf,k+1|k}^{(i)} \mathbf{H}_f^{(i)^\top} + \mathbf{H}_f^{(i)} \mathbf{P}_{f,k+1|k}^{(i)} \mathbf{H}_f^{(i)^\top} + \mathbf{R}_{k+1}^{(i)} \tag{1.42}$$

To include the landmark correlation information in the test of measurement and map pairings, one must recur to the joint compatibility test.

Joint compatibility test

Landmark correlation is taken into account when we test spatial match hypotheses for more than one landmark at a time. The joint measurement innovation covariance for two or more landmarks is

$$\mathbf{S}^{(p\ldots r)} = \mathbf{H}^{(p\ldots r)} \mathbf{P}_{k+1|k} \mathbf{H}^{(p\ldots r)^\top} + \mathbf{R}_{k+1}^{(p\ldots r)} \tag{1.43}$$

or equivalently,

$$\mathbf{S}^{(p\ldots r)} = \begin{bmatrix} \mathbf{S}^{(p\ldots q)} & \mathbf{H}^{(p\ldots q)} \mathbf{P}_{k+1|k} \mathbf{H}^{(r)^\top} \\ \mathbf{H}^{(r)} \mathbf{P}_{k+1|k} \mathbf{H}^{(p\ldots q)^\top} & \mathbf{S}^{(r)} \end{bmatrix} \tag{1.44}$$

The indices $(p \ldots r)$ need not be consecutive, with o the total number of landmarks in the set $(p \ldots r)$. The joint Mahalanobis distance is

$$d_{p\ldots r}^2 = \tilde{\mathbf{z}}_{k+1|k}^{(p\ldots r)\top} \mathbf{S}^{(p\ldots r)^{-1}} \tilde{\mathbf{z}}_{k+1|k}^{(p\ldots r)} \tag{1.45}$$

and the joint compatibility test for two or more landmarks is

$$d_{p\ldots r}^2 \leq \chi_{do,\alpha}^2 \tag{1.46}$$

The joint compatibility test can be computationally expensive when the number of landmarks involved is large. It would require the inverse of a joint measurement covariance matrix of size $do \times do$ with complexity $O((do)^3)$. This complexity can be reduced however, as shown next, with an *incremental* computation of $\mathbf{S}^{(p\ldots r)^{-1}}$.

Without loss of generality, once the inverse up to the q-th landmark $\mathbf{S}^{(p\ldots q)^{-1}}$ has been computed, from the matrix inverse lemma for block matrices[1], we have that

[1] See Appendix B.

$$\mathbf{S}^{(p\dots r)^{-1}} = \begin{bmatrix} \mathbf{S}^{(p\dots q)^{-1}} + \mathbf{S}^{(p\dots q)^{-1}}\mathbf{M}\mathbf{N}\mathbf{M}^{\top}\mathbf{S}^{(p\dots q)^{-1}} & \mathbf{S}^{(r)^{-1}}\mathbf{M}\mathbf{N} \\ -\mathbf{N}\mathbf{M}^{\top}\mathbf{S}^{(r)^{-1}} & \mathbf{N} \end{bmatrix} \tag{1.47}$$

with

$$\mathbf{M} = \mathbf{H}_{\mathbf{x}}^{(p\dots q)}\mathbf{P}_{k+1|k}\mathbf{H}_{\mathbf{x}}^{(r)^{\top}} \tag{1.48}$$

$$\mathbf{N} = (\mathbf{S}^{(r)^{-1}} - \mathbf{M}^{\top}\mathbf{S}^{(r)^{-1}}\mathbf{M})^{-1} \tag{1.49}$$

At the end, the computation of the full inverse $\mathbf{S}^{(p\dots r)^{-1}}$ is simplified with an incremental computation requiring a total of o individual covariance matrices as indicated in Equation 1.49. Values of $\chi^2_{do,\alpha}$ for $\alpha = 0.95$ and $\alpha = 0.99$, are tabulated in Tables 1.1 and 1.2 respectively.

Table 1.1. The Pearson χ^2 distribution needed for the computation of the square Mahalanobis distance threshold $d^2 = \chi^2_{do,\alpha}$ during the validation of the hypothesis test that measurements $\mathbf{z}^{(p\dots r)}$ correspond to features $\mathbf{x}_f^{(p\dots r)}$ with probability $\alpha = 0.95$.

Feature model	points on a line	image points, lines on a plane	points, lines, planes in 3d
d	1	2	3
Number of features o	$\chi^2_{o,0.95}$	$\chi^2_{2o,0.95}$	$\chi^2_{3o,0.95}$
1	3.84	5.99	7.91
2	5.99	9.49	12.6
3	7.81	12.6	16.9
4	9.49	15.5	21.0
5	11.1	18.3	25.0
10	18.3	31.4	43.8
15	25.0	43.8	61.6
20	31.4	55.8	79.1
30	43.8	79.1	113.1

If the regions of uncertainty around each landmark in the map do not overlap, the individual compatibility test (ICT in the sequel) might be sufficient to obtain a robust solution to the data association problem in SLAM. Such is the case of sparse maps, or when precise sensors are available.

However, in situations with large amounts of clutter, or with very uncertain sensor models, the joint compatibility test (JCT in the sequel) might be more useful. One should not take for granted however, the increase in the computational cost incurred by the JCT. Testing the joint compatibility of

Table 1.2. The Pearson χ^2 distribution needed for the computation of the square Mahalanobis distance threshold $d^2 = \chi^2_{do,\alpha}$ during the validation of the hypothesis test that measurements $\mathbf{z}^{(p\ldots r)}$ correspond to features $\mathbf{x}_f^{(p\ldots r)}$ with probability $\alpha = 0.99$.

Feature model	points on a line	image points, lines on a plane	points, lines, planes in 3d
d	1	2	3
Number of features o	$\chi^2_{o,0.99}$	$\chi^2_{2o,0.99}$	$\chi^2_{3o,0.99}$
1	6.63	9.21	11.3
2	9.21	13.3	16.8
3	11.3	16.8	21.7
4	13.3	20.1	26.2
5	15.1	23.2	30.6
10	23.2	37.6	50.9
15	30.6	50.9	70.0
20	37.6	63.7	88.4
30	50.9	88.4	124.1

tens of landmarks might deem impossible a real time implementation of the algorithm.

In situations where multiple data association hypotheses are to be verified, the joint compatibility criterion to landmark matching is implemented in branch and bound algorithms that search efficiently the space of compatible landmark association solutions. An exemplary application of one of such heuristics is detailed in the aforementioned contribution by Neira and Tardós [70].

Figure 1.5 exemplifies the use of the ICT and JCT for data association hypothesis validation. The left column of plots in the Figure shows a robot localization sequence with three landmarks. For such a small number of landmark observations, the number of match hypothesis that fail the test is similar in both cases (6.6% for ICT and 8% for JCT). The localization error results of using the ICT and JCT in this case are nearly equivalent.

On the other hand, the right column of plots in the same Figure shows the results of using the ICT and JCT in a localization sequence with 10 landmarks and similar noise parameters. In this case, the number of times that match hypothesis fail the JCT nearly doubles those of the ICT (30% for JCT vs. 17.7% for ICT); indicating perhaps that the ICT was overconfident when testing for data association. Landmark observations do not pass the joint compatibility test unless data association is accurate for all observations, or until the uncertainty $\mathbf{P}_r$ has grown to such value as to make the all-landmark

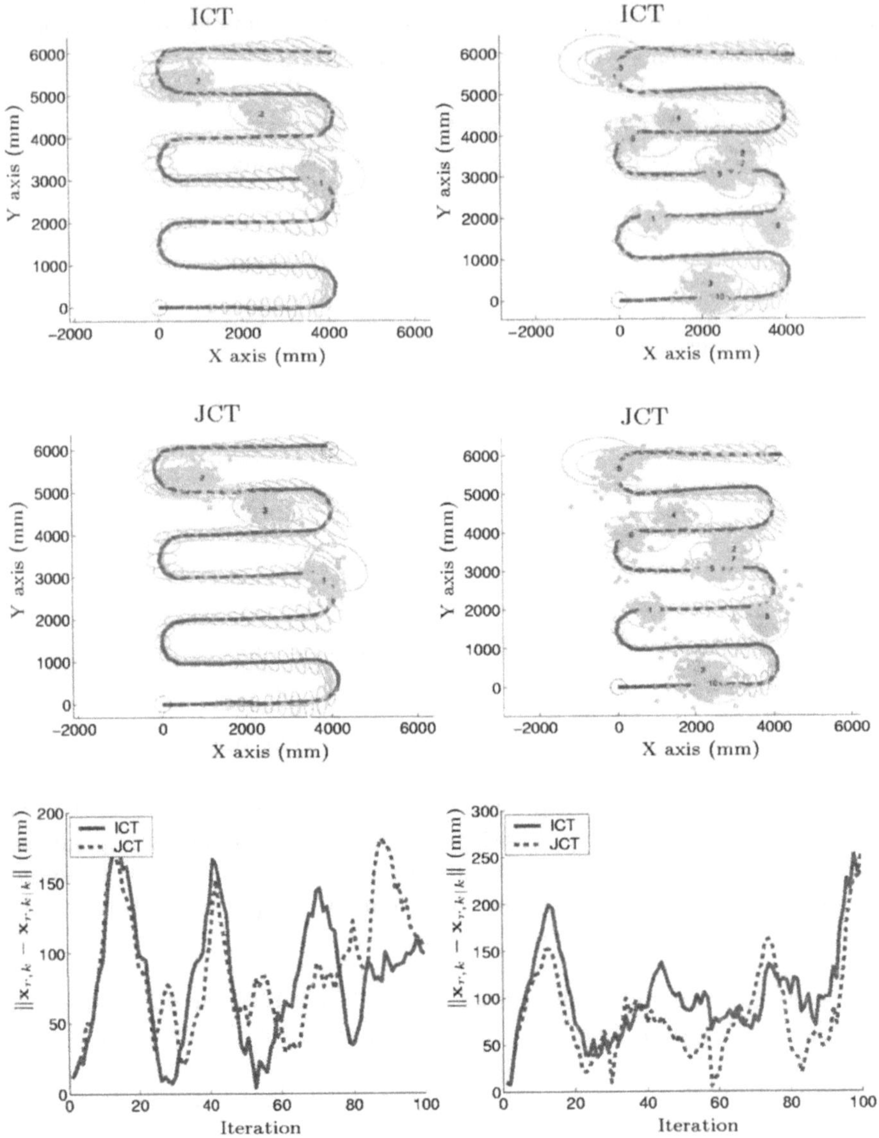

a) 3 landmarks

b) 10 landmarks

Fig. 1.5. Spatial compatibility tests. Individual and joint compatibility test performance for a map with 3 and 10 landmarks over 100 steps.

Fig. 1.6. *Monobot*, a 1-D mobile robot.

Mahalanobis distance $\tilde{\mathbf{z}}\mathbf{S}^{-1}\tilde{\mathbf{z}}$ sufficiently small. It might happen in the JCT case, that for a large number of the filter iterations (30% in this test run), vehicle location estimates end up being computed from dead reckoning only. It is certainly better however not to revise the robot localization at all, than doing so with incorrect data association. As indicated in the right bottom plot, the overall performance of the algorithm shows some improvement when the JCT is used.

The dots in the ICT map plots represent all of the observations that pass the test, whereas the dots in the JCT map plots show all landmark observations, regardless if they passed the test or not. The estimated robot localization error is represented by the ellipses along the trajectory, which correspond to 2σ level curves of the robot pose error covariance estimate. The ellipses around the landmarks on the other hand, indicate projected 2σ bounds for the landmark covariance estimates. The solid line and the dashed line in the two bottom plots are used to indicate ICT and JCT localization error, respectively.

1.2 Mobile Robot Platforms

In this section we provide explicit expressions for two mobile robot configurations, a one-dimensional (1-D) robot moving along a straight line, and a planar wheeled mobile robot on a two-dimensional (2-D) environment.

Monobot

Consider the one-dimensional robot (*monobot*) from Figure 1.6. The robot location is $x_{r,k}$, and the motion command is u_k. The robot error dynamics is v_k, and the vehicle process model is simply

$$x_{r,k+1} = x_{r,k} + u_k + v_k \tag{1.50}$$

The map consists of the set $\mathbf{x}_f$ of static landmark points. The measurement equation for each landmark is

$$z_k^{(i)} = x_f^{(i)} - x_{r,k} - w_k^{(i)} \tag{1.51}$$

with $w_k^{(i)}$ the landmark observation error.

The study of this simple robot model might seem naive at this point. Its relevance will become evident in the next Chapter, where we study in detail the observability and controllability aspects of the stochastic approach to SLAM. Suffice to say for the time being, that thanks to the linearity of the model, the approximations in Equations 1.6 and 1.7 are exact, and that

$$\mathbf{F} = \mathbf{G} = \mathbf{H}_f = 1 \tag{1.52}$$

$$\mathbf{H}_r = -1 \tag{1.53}$$

Planar mobile robot

We now develop expressions for the more realistic planar robot shown in Figure 1.7, a nonlinear nonholonomic velocity controlled wheeled vehicle with three degrees of freedom, and an environment consisting of 2-D point landmarks located on the floor.

The vehicle is controlled by a linear velocity v and a steering velocity ω. The process model used to predict the trajectory of the center of projection of the laser range scanner is given by

$$\begin{bmatrix} x_{k+1} \\ y_{k+1} \\ \theta_{k+1} \end{bmatrix} = \begin{bmatrix} x_k + \tau\left((v_k + v_{v,k})\cos\theta_k - (\omega_k + v_{\omega,k})l\sin\theta\right) \\ y_k + \tau\left((v_k + v_{v,k})\sin\theta_k + (\omega_k + v_{\omega,k})l\cos\theta\right) \\ \theta_k + \tau(\omega_k + v_{\omega,k}) \end{bmatrix} \tag{1.54}$$

where l is the distance from the center of the wheel axle to the center of projection of the laser range scanner, τ is the time constant, and v_v, v_ω are zero mean Gaussian model noises.

The observation model is

$$\begin{bmatrix} z_{r,k}^i \\ z_{\beta,k}^i \end{bmatrix} = \begin{bmatrix} \sqrt{(x_f^i - x_k)^2 + (y_f^i - y_k)^2} + w_{r,k} \\ \tan^{-1}\left(\frac{(y_f^i - y_k)}{(x_f^i - x_k)}\right) - \theta_k + \frac{\pi}{2} + w_{\beta,k} \end{bmatrix} \tag{1.55}$$

with z_r^i and z_β^i the distance and bearing of an observed point landmark with respect to the laser center of projection. x_f^i and y_f^i are the absolute coordinates of such landmark, and i is used for the labelling of landmarks.

The Jacobian matrices $\mathbf{F}$, $\mathbf{G}$, and $\mathbf{H}$ are obtained by differentiating Equations (1.54) and (1.55) with respect to states and noises. That is,

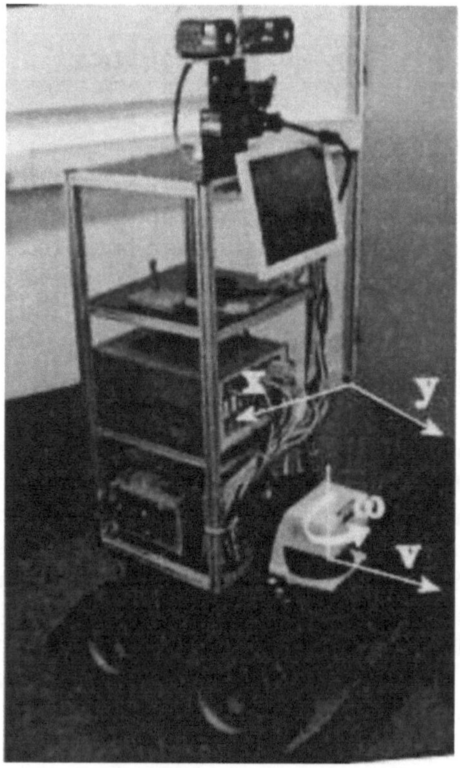

Fig. 1.7. Two-dimensional mobile robot.

$$\mathbf{F} = \begin{bmatrix} 1\ 0 & -\tau((v_k + v_{v,k})\sin\theta_k - l\cos\theta(\omega_k + v_{\omega,k})) & 0\ 0 & \\ 0\ 1 & \tau((v_k + v_{v,k})\cos\theta_k - l\sin\theta(\omega_k + v_{\omega,k})) & 0\ 0 & \\ 0\ 0 & 1 & 0\ 0 & \\ & & \ddots & \\ 0\ 0 & 0 & 1\ 0 & \\ 0\ 0 & 0 & 0\ 1 & \\ & & & \ddots \end{bmatrix} \quad (1.56)$$

$$\mathbf{G} = \begin{bmatrix} \tau\cos\theta_k & -\tau l\sin\theta_k \\ \tau\sin\theta_k & +\tau l\cos\theta_k \\ 0 & 1 \\ \vdots & \\ 0 & 0 \\ 0 & 0 \\ \vdots & \end{bmatrix} \quad (1.57)$$

$$\mathbf{H}_i = \begin{bmatrix} -\frac{x_f^{(i)}-x_k}{d} & -\frac{y_f^{(i)}-y_k}{d} & 0 & \cdots & \frac{x_f^{(i)}-x_k}{d} & \frac{y_f^{(i)}-y_k}{d} & \cdots \\ \frac{y_f^{(i)}-y_k}{d^2} & -\frac{x_f^{(i)}-x_k}{d^2} & -1 & & -\frac{y_f^{(i)}-y_k}{d^2} & \frac{x_f^{(i)}-x_k}{d^2} & \end{bmatrix} \tag{1.58}$$

with $d = \sqrt{(x_f^{(i)} - x_k)^2 + (y_f^{(i)} - y_k)^2}$.

As a motivational note to the reader, this detailed formulation of the kinematics of the *monobot* and the planar robot are to be used in Chapter 2 in which we show how the limitations inherent in the reconstructibility of the map state space are equally present in both vehicle models.

1.3 Temporal Landmark Validation

Now that we have underpinned the fundamental characteristics the original EKF-SLAM algorithm, as well as the specifics of the mobile robot platforms used, we turn our attention to the study of appropriate temporal landmark quality functions that help in the alleviation of the data association problem.

We reviewed in Sections 1.1-1.1 the *spatial* landmark quality tests already present in the literature [70]. We introduce, in Sections 1.3-1.3 two new *temporal* landmark quality models, and show the feasibility of using these in a test to validate not only where, but when should we expect future observations for any given landmark.

Landmark temporal uncertainty

As indicated in the introductory section to this Chapter, the use of spatial compatibility tests is crucial for the solution of data association in SLAM, but they can still be insufficient in situations with moderate scene dynamics. Imagine that a spurious landmark observation was sufficiently robust to be added to the map. This landmark could come from a temporary artifact in the scene such as an open door, or an artifact in our sensor such as a shadow or reflection that is persistent only during a small number of iterations.

With the static map model presented so far, once the spurious landmark disappears from the scene, or at least, from the sensor return, we have no means to revise the map and delete it. The two spatial compatibility tests presented would not have any information on the history of observations, and would be trying to associate new observations to that entry in the map. If the algorithm succeeds at incorrectly associating new landmark observations with such previously stored weak feature, i.e., the observation fell inside the uncertainty ellipse described by $\mathbf{S}$ and passed the data association test; it will most likely introduce landmark localization error in the map, while at the same time decreasing the estimated map error covariance. Given that the map covariance is fully correlated, starting with the next iteration of the algorithm, that uncertainty would be propagated to the rest of the landmark locations, and that of the robot as well; ultimately breaking down the entire estimation approach to SLAM.

For moderately dynamic environments we need some means to erase landmarks from the map once we have realized they are not sufficiently robust, not only during landmark initialization, but during the entire run of the algorithm. To aid in those situations in which the landmark observations might not be consistent in time, we propose a new set of landmark quality functions. The idea is to use these functions as an aid to validate temporal landmark quality.

The first temporal landmark quality function proposed is an exponential decay rule used to learn the persistence of landmark matching. The second function is a linear model used to update the probability of data association from the sequence of landmark matches. These two functions are used to validate landmark observations in time, and ultimately to erase from the map those landmarks that are not sufficiently persistence in the scene. In Section 1.3 we analyze the consequences of removing landmarks from the map with respect to the evolution of the fully correlated covariance entries in $\mathbf{P}$.

Nonlinear model for temporal landmark quality: the exponential decay rule

One possibility in the computation of the temporal landmark quality is to have an exponential decay rule. This way, each landmark in the map will have an associated memory cell that registers how persistent, and how old that landmark is.

Imagine that at the $k+1$-th iteration, the i-th landmark measurement estimate $\mathbf{z}^{(i)}_{k+1|k}$ falls inside the current field of view, but none of the entries in the observation vector $\mathbf{z}_{k+1}$ has similar appearance properties, nor is sufficiently close (in the sense of d^2) to pass the spatial landmark compatibility tests. This would be the situation if, for example, the i-th landmark was learned from a temporary artifact in the scene that was only tracked over sensor data for a short period of time, but is no longer present. With the aid of an exponential decay rule to data association, its quality measure will decay in the absence of observation matches, indicating the map building algorithm that such landmark is no longer present in the scene and should not be considered a relevant feature for robot localization.

We propose a nonlinear update rule for landmark quality of the form

$$x^{(i)}_{q,k+1} = \frac{1}{1 + e^{-\left(\alpha u^{(i)}_{q,k} + \beta x^{(i)}_{q,k}\right)}} \tag{1.59}$$

where $u^{(i)}_{q,k}$ is the landmark identification stamp

$$u^{(i)}_{q,k} = \begin{cases} 0 & : \text{ failed the spatial data association test} \\ 1 & : \text{ passed the spatial data association test} \end{cases} \tag{1.60}$$

The scalar α is an input weight used to regulate the contribution of such landmark identification over the previous map configuration, and β is a memory weight used to regulate the contribution of the previous landmark quality state over its new value.

The asymptotic lower and upper bounds of Equation 1.59 can be evaluated by solving the equations

$$x_{q,LOW} = \frac{1}{1+e^{-(\beta x_{q,LOW})}} \tag{1.61}$$

$$x_{q,HIGH} = \frac{1}{1+e^{-(1+\beta x_{q,HIGH})}} \tag{1.62}$$

Using a symbolic manipulation math package, we find for example, that for $\alpha = \beta = 1$, $x_{q,LOW} = 0.6590$, and $x_{q,HIGH} = 0.8659$. Landmark initialization is at the middle of the scale, i.e, $x^{(i)}_{q,0} = 0.7682$.

Linear model for temporal landmark quality: the data association probability

Another possibility in the computation of the temporal landmark quality is to consider the probability of correct data association of such landmark in the next iteration.

According to the relative frequency definition of probability, if an event (say, the correct association of landmark i) occurs j times in k trials (observations), and provided k is sufficiently large, then the probability that the same landmark will be properly matched in the next iteration can be expressed as

$$p^{(i)}_k = \frac{j}{k} \tag{1.63}$$

Now, once a new observation is made, the data association probability will change according to the new landmark association result. This change in probability is represented by the recursive expression

$$p^{(i)}_{k+1} = \frac{p^{(i)}_k k + u^{(i)}_{q,k}}{k+1} \tag{1.64}$$

with $u^{(i)}_{q,k}$ defined as in Equation 1.60. If we make the notation change $a = k/(k+1)$, and $x^{(i)}_{q,k} = p^{(i)}_k$, our second model for temporal landmark quality becomes

$$x^{(i)}_{q,k+1} = a x^{(i)}_{q,k+1} + (1-a) u^{(i)}_{q,k} \tag{1.65}$$

For fixed values of k, the constant a accounts for a memory weight with a role similar as those of α and β from the previously discussed model. It can be fixed to a constant value between 0 and 1, and it indicates the memory length to be used in the computation of the new data association probability. So

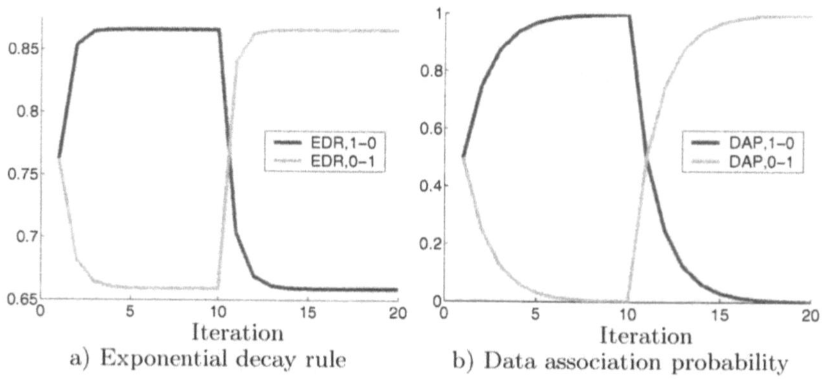

Fig. 1.8. Landmark quality models.

for example, if a memory window of the last 5 iterations is to be considered, the memory weight becomes $a = 0.8333$. In this linear model for temporal landmark quality, x_q is bounded between 0 and 1, and initialization is made at 0.5.

Both temporal landmark quality measures, the exponential decay rule, and the data association probability were chosen to have similar properties. For instance, to be asymptotically bounded by above and below by

$$x_{q,LOW} \leq x_q \leq x_{q,HIGH} \tag{1.66}$$

Any other function with such monotonicity could be also used as temporal landmark quality function. However, such function must have some way of tuning the memory length of the algorithm. The two functions proposed do so by means of the parameters α, β, and a. The left plot in Figure 1.8 shows the behavior of the exponential decay rule for the parameter values $\alpha = 1$, and $\beta = 1$. Similarly, the right plot shows the data association probability with parameter $a = 0.5$. The labels 0-1 and 1-0 indicate the change in the landmark identification stamp u_q from 0 to 1, and from 1 to 0, representing the presence or loss of data association.

Temporal landmark quality test

In the same way that the spatial compatibility test is used to validate if observations are consistent with the already learned map entries; the temporal landmark quality test must be used to validate if any map entry is sufficiently robust to be kept in the map. The test verifies if the history of data association has kept the value for the temporal landmark quality above a user defined cut threshold q. All landmarks expected to appear in the current field of view, and for which no occlusion has been predicted, must have their landmark quality

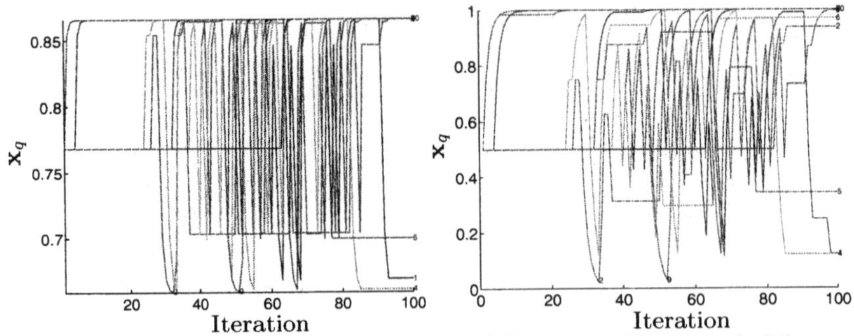

a) Exponential decay rule   b) Data association probability

Fig. 1.9. Landmark quality test for a test run with 10 landmarks and 100 steps. $\alpha = 1$, $\beta = 1$, and $a = 0.5$. The plots show landmarks labeled 2 and 9 being reinitialized near steps 30 and 50 respectively.

measure updated. Furthermore, those landmarks whose temporal quality measure falls below the user defined threshold should be removed from the map. The heuristics needed to handle occlusions, depend on the type of landmarks and sensors used.

The temporal landmark quality test is

$$\textbf{if } x_q^{(i)} \leq q$$
$$\texttt{RemoveLandmark}(\mathbf{x}_f^{(i)})$$

In the case of the exponential decay rule with parameters $\alpha = 1$ and $\beta = 1$, for example, the cut threshold $q = 0.66$ is reached once a landmark has not been observed for 5 consecutive iterations, or more if these were not consecutive. Similar effects are obtained when using the data association probability with the parameter value $a = 0.5$, and a cut threshold of $q = 0.03$. Figure 1.9 shows both the exponential decay rule, and the data association probability as landmark quality measures for a test run of 100 steps and 10 landmarks, with slightly modified simulation parameters than the ones used to create Figure 1.5b. In this test run, we have forced 25% of the observations to be misidentified to their closest neighbor. The individual compatibility test catches some but not all of these mismatches, and yields an identification stamp value $u_q = 0$ for them. By adding the more restrictive temporal landmark quality tests, those landmarks with a large amount of mismatches end up being removed from the map, and are reinitialized as new landmarks once they become robust again.

We have opted for a simplified heuristic for the removal of a landmark from the map, with the advantage of computational efficiency, but at the expense of suboptimality. Our algorithm simply erases the low quality landmark entries from the state vector and its corresponding row and column in the state error covariance matrix. Once the landmark is robust again, it is considered as a new different landmark, and initialized according to the discussion from Section 1.1. Note however that, given the fact that SLAM is fully correlated, the contribution of a misidentified landmark estimate in revising the error covariance matrix has already propagated to the entire map. The right thing to do, would be to trace back the intermediate results of the algorithm up to the point in which the landmark was originally inserted, and to recompute forward once more the map state and map error covariance up to the current iteration, without considering that landmark, as if it had never existed.

Saving the state vector and error covariance matrix for all iterations has space complexity of $O(kn^2)$, with k the number of iterations, and n the number of landmarks. Furthermore, recomputing the state vector and error covariance matrix from the point at which the spurious landmark was initially inserted, would most likely lead to different values on $\mathbf{P}$, and consequently on $\mathbf{S}$, producing even different data association results. So, not only the state vector and covariance matrix history must be maintained, but the full measurement data as well, requiring for a full run of the algorithm every time a landmark is found to be spurious. The optimal solution is rather cumbersome, and we have opted for suboptimality, with the aforementioned simplification of just deleting the corresponding entries in $\mathbf{x}$ and $\mathbf{P}$, with the following insight.

Gibbens *et. al.* [41], show how in SLAM all entries in the covariance matrix $\mathbf{P}$ depend on the number of landmarks used in the form of the total Fisher information I_T. This is a measure of the total information per unit time available to the filter. For a monobot with n landmarks, all with equal measurement variance σ_w^2, the total Fisher information is $I_T = n/\sigma_w^2$. The more landmarks available, the more information the filter has. That is, the greater the number of landmarks used, the smaller the asymptotic values for the entries in the error covariance.

The removal of a landmark from the state vector in the form discussed is consistent with this observation. We next show, by example, how for the monobot, all the entries in $\mathbf{P}$ with the removal of a landmark at some point in the algorithm are bounded by below and above by the same entries in $\mathbf{P}$, but with and without considering the landmark for the entire run. Let us call $\mathbf{P}_{\cdot,n}$ the entry in $\mathbf{P}$ for a map with n landmarks, and $\mathbf{P}_{\cdot,n+1,n}$ the entry in $\mathbf{P}$ for a map that went from $n+1$ to n landmarks via the removal of landmark states. Then, for the entire run of the algorithm

$$\mathbf{P}_{r,k|k,n+1} \leq \mathbf{P}_{r,k|k,n+1,n} \leq \mathbf{P}_{r,k|k,n} \tag{1.67}$$

$$\mathbf{P}^{(i)}_{rf,k|k,n+1} \leq \mathbf{P}^{(i)}_{rf,k|k,n+1,n} \leq \mathbf{P}^{(i)}_{rf,k|k,n} \tag{1.68}$$

$$\mathbf{P}^{(i,j)}_{f,k|k,n+1} \leq \mathbf{P}^{(i,j)}_{f,k|k,n+1,n} \leq \mathbf{P}^{(i,j)}_{f,k|k,n} \tag{1.69}$$

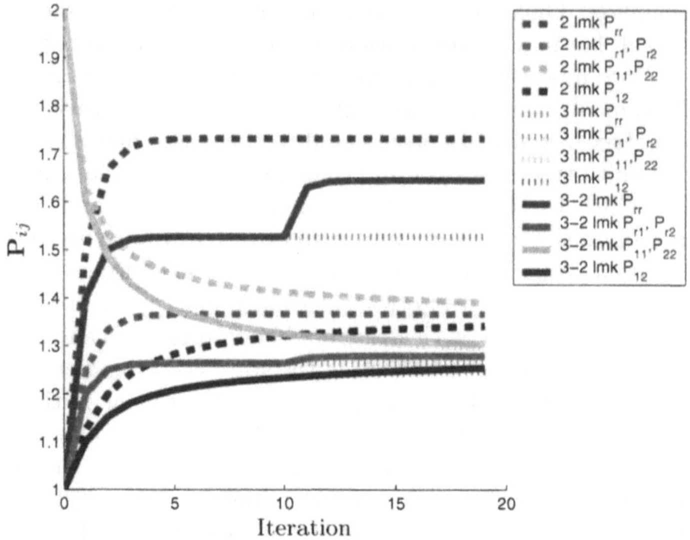

Fig. 1.10. Evolution of the entries in the covariance matrix for a monobot with 3, 2, and 3-then-2 landmarks.

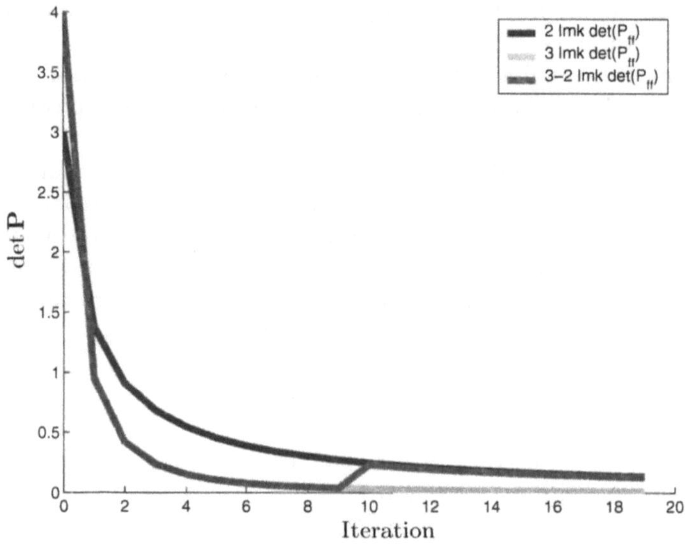

Fig. 1.11. Asymptotic convergence of the determinant of the map error covariance matrix for a monobot with 3, 2, and 3-then-2 landmarks.

Moreover, by removing a landmark from the map in the form discussed, the asymptotic convergence property from Equation 1.25 is maintained. That is, the revised map is still fully correlated. Figure 1.10 shows the evolution of the entries in the covariance matrix for a monobot with two and three landmarks and noise parameters $\mathbf{Q} = 1$, $\mathbf{R}^{(i)} = 1$, $\mathbf{P}_{r,0|0} = 1$. Figure 1.11 shows the evolution of the determinant of the map error covariance for the same monobot configuration, an indication of asymptotic full correlation.

1.4 Performance of EKF SLAM with Landmark Validation

Having underpinned the characteristics and model assumptions of the full covariance EKF-SLAM and EKF-SLAM-LV algorithms (LV stands for the landmark validation extension to the original algorithm), we present now results on the improvement in the reconstruction achieved. Namely, we will compare the original algorithm with the modifications induced by sequential innovation and observation range limitations. We will then concentrate on the more realistic limitations produced by erroneous data association, which is considered one of the most critical artifacts that might destroy the viability of the EKF-based method to simultaneous localization and map building. This is, when one or more observed landmarks are misidentified, their location estimate might diverge considerably, consequently inducing large errors in the localization of the robot as well. We will show how this problem is alleviated by using both temporal and spatial landmark quality tests to verify the correctness of data association.

Sequential innovation

From this point on we will consider as our standard test case, and unless otherwise indicated, a planar robot traversing an environment with 10 landmarks in 100 steps, with 10% standard deviation over desired polar motion commands (rotational and translational velocities) to simulate dead reckoning, and with 1% measurement accuracy, also in polar coordinates. Figure 1.12 shows the results of applying the full-covariance EKF-SLAM algorithm, and the computationally more efficient version with sequential innovation. In the first two plots in the figure, the continuous line indicates the actual robot displacement, and the continuous line with small ellipses on top, representing the estimated trajectory. The ellipses are projected uncertainties at a distance of 2σ. Furthermore, the sparse light dots indicate landmark measurements, and the darker dots indicate their location estimate as computed by the localization algorithm.

As the reader can appreciate in the error plots, both implementations are practically equivalent, the implementation using sequential innovation being slightly better. The actual difference in the localization of the mobile robot for

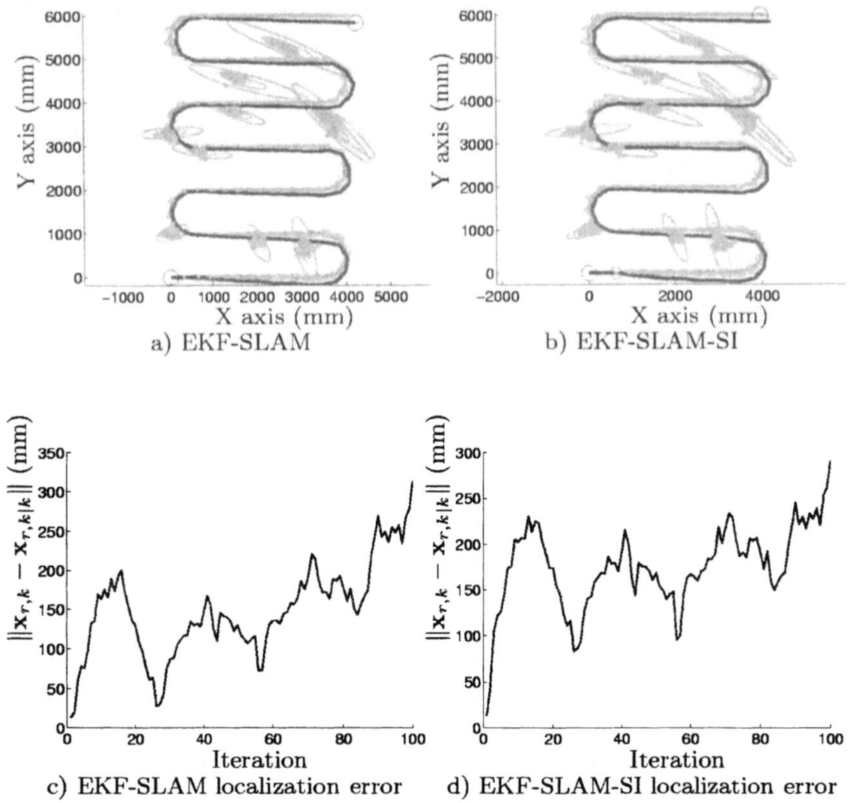

Fig. 1.12. Full-covariance EKF SLAM for a path with 100 iterations and 10 landmarks.

both implementations is shown in Figure 1.13, with a variation in the localization estimate by $1.68mm$ on average, for a run of nearly $40m$, a deviation of less than 0.0042%, and for all practical purposes negligible. Given that the two methods are theoretically equivalent, this small divergence is attributed to floating point computations and quantization, and to the better model estimates obtained when re-computing the model Jacobians after each measurement is assimilated by the filter. In order to appreciate the contribution in localization error from each degree of freedom of the vehicle, Figure 1.14 shows the components of robot localization error in the EKF-SLAM run with sequential innovation, with 2σ bounds signaled as dotted lines.

Figures 1.15 and 1.16 help visualize the behavior of the entries in the error covariance matrix for the EKF-SLAM algorithm with sequential innovation. Figure 1.15 shows for example how the uncertainty in the location of the landmarks reduces monotonically, with each line representing the landmark noise covariance estimate in both the x and y axes.

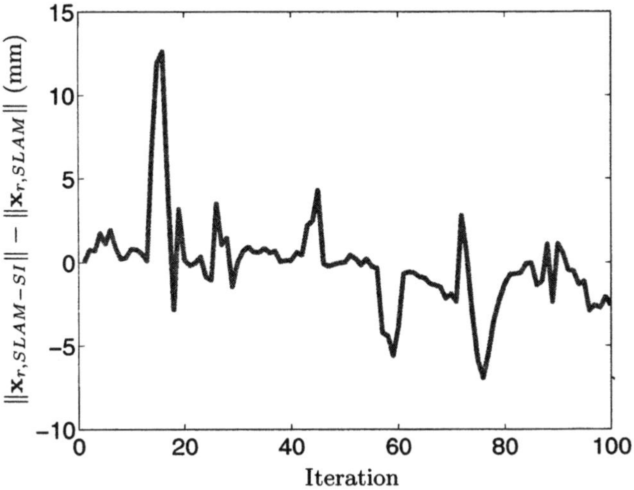

Fig. 1.13. Difference in localization error between the full covariance EKF version of the algorithm, and the results of applying sequential innovation to the same set of observations.

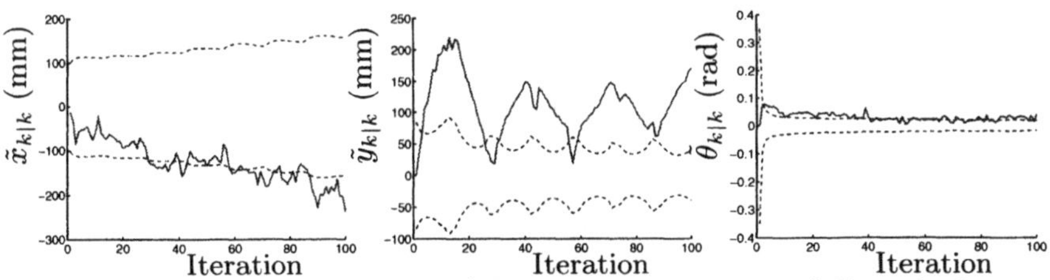

Fig. 1.14. Robot localization error estimate. EKF SLAM with sequential innovation for a path with 100 iterations and 10 landmarks.

One direct consequence from Equation 1.25 is that the correlation coefficients of the map covariance $\mathbf{P}_f$ are monotonically increasing. So at each iteration, these correlation coefficients can be evaluated with

$$\rho_{x,k}^{(ij)} = \frac{\mathbf{P}_{f_x,k|k}^{(ij)}}{\sqrt{\mathbf{P}_{f_x,k|k}^{(ii)} \mathbf{P}_{f_x,k|k}^{(jj)}}} \tag{1.70}$$

$$\rho_{y,k}^{(ij)} = \frac{\mathbf{P}_{f_y,k|k}^{(ij)}}{\sqrt{\mathbf{P}_{f_y,k|k}^{(ii)} \mathbf{P}_{f_y,k|k}^{(jj)}}} \tag{1.71}$$

and we show in Figure 1.16 the evolution in time of the correlation coefficients for one such landmark with respect to the rest.

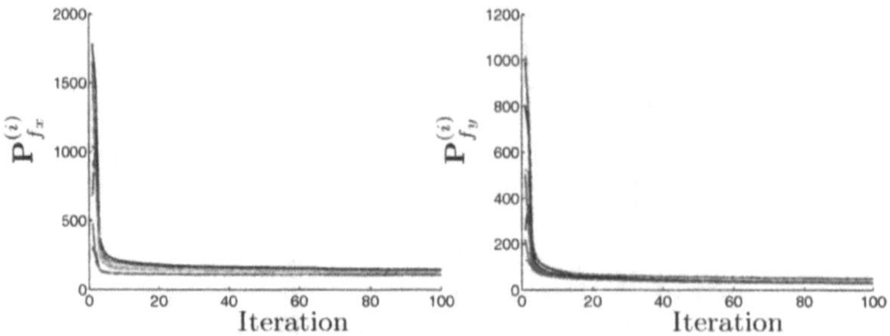

a) Covariance in x b) Covariance in y

Fig. 1.15. The landmark noise covariance estimate is monotonically decreasing in the EKF SLAM algorithm.

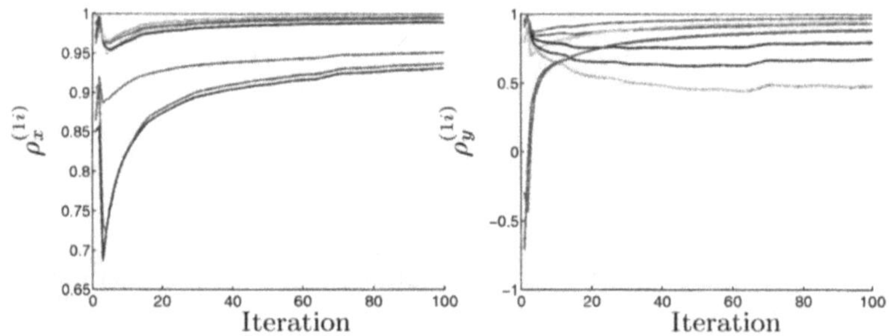

a) Cross correlation in x b) Cross correlation in y

Fig. 1.16. Landmarks become more and more correlated as iterations take place in the EKF SLAM algorithm. The figure shows the increase in the cross correlation coefficients for landmark 1 with respect to each other landmark, in both the x and y directions.

Partial observations

Now, imagine a more realistic sensor, one with a limited field of view. Will the fact that not all landmarks can be seen and matched at all times worsen the reconstruction results? The answer is twofold. One the one side, and as will be seen in the next Chapter, when using a reduced number of landmarks, the effects of partial observability are accentuated, limiting the reconstruction of the vehicle and landmark locations. On the other hand, given the measurement noise models proposed, ones that are proportional to the distance to the observed landmark; by neglecting measurements from objects that are far from the robot, we also limit the value of the noise introduced by the measurement model.

These arguments might lead to the formulation of a strategy for the computation of the appropriate number of landmarks to use, optimal in the sense of the reconstruction results (reduced partial observability effects), and at the same time limiting the impact of sensor measurement noise (by eliminating unreliable observations in terms of the distance of the landmarks to the robot).

Let us analyze in greater detail, and with the aid of our standard test case from Section 1.4, how a limited field of view undermines the accuracy in localization when using the EKF-SLAM algorithm. Figure 1.17 shows a run of the algorithm with sequential innovation for a sensor with a limited field of view of $2m$. Notice how in between the 5th and 15th iterations approximately, the robot pose estimate depends on observations from 2 landmarks only. The growth in size of the covariance ellipses clearly shows the increase in the uncertainty in localization. By the time new landmark observations enter the field of view, the vehicle has already accumulated a large amount of dead reckoning error. Consequently, all initial landmark position and covariance estimates are biased with such initial localization error.

A similar situation occurs near the end of the simulation, around the 85th iteration. In this case, the map is revised with repeated observations of just one landmark only, and even when the localization estimate remains consistent with such observation, just one 2d landmark does not suffice in localizing a 3dof vehicle. The accumulated deviation is corrected once a previously learned landmark re-enters the field of view.

Consequently, the two most serious artifacts that can hinder accurate localization when the sensors have a limited field of view are, the wrong initialization of landmark estimates due to accumulated error, and the decrease in the dimensionality of the observed references during motion, up to the point in which pose recovery becomes ill-posed.

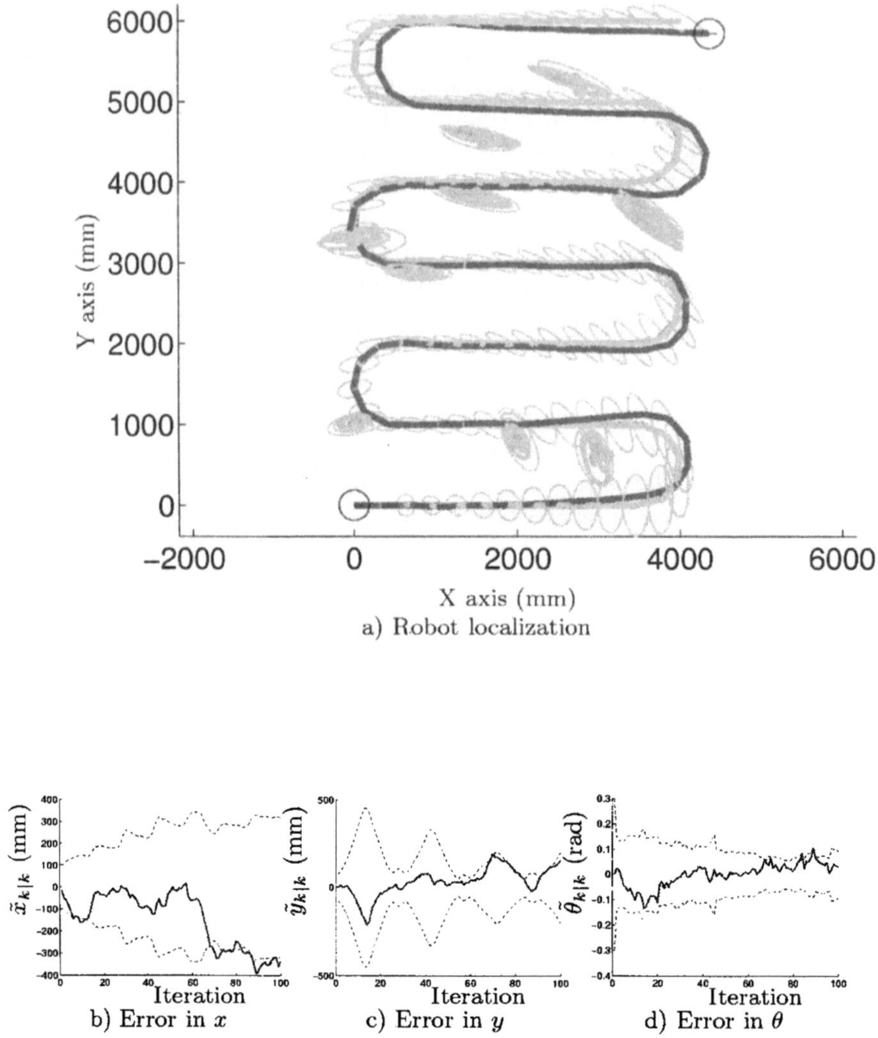

Fig. 1.17. Full-covariance EKF CML for a path with 100 iterations and 10 landmarks, and a sensor with a limited radius of observation of $2m$.

Data missassociation

A more drastic situation is the case when the landmark identification module is not error-free. That is the case when we allow the system a percentage of landmark identification mismatches. Imagine that we observe say visual landmarks, such as corners or lines extracted from intensity images, and that our landmark tracking algorithm is not very accurate at matching observations in consecutive frames because of illumination changes or other data association artifacts. To simulate this behavior we rely once more on our standard test case. For the sample run shown in Figure 1.18, landmark matching is performed with a 25 percent probability of missassociation within a $1m$ radius.

The effects of landmark misidentification are much more pervasive than the lack of observations explained in the previous section. In the sample run shown in Figure 1.18, localization and map building proceeds smoothly until the 30th iteration, when the first mismatch occurs. The algorithm is not able to recover from this failure, and when the previously observed landmarks re-enter the field of view two things happen. On the one hand, the new observations of the already learned landmarks aid in reversing the error trend in localization. On the other hand, the same new observations are used to revise the mere location estimates of those landmarks. Moreover, new landmark observations will be initialized with corrupt robot location estimates.

Given the fact that the map is fully correlated, the effects of landmark mismatch propagate to the localization estimate of all the landmarks in the model.

The estimation theoretic approach to SLAM, as presented by Smith and Cheeseman, and later formalized by Leonard, Newman, and Durrant-Whyte among others, is very sensitive to data association errors; and as formulated lacks a theoretical foundation to deal with the problem. Efforts have been tailored at correcting the effects of data mismatches, and at finding measures of the spatial compatibility of landmark correspondence. We go one step further, providing a new formulation of temporal landmark quality measures.

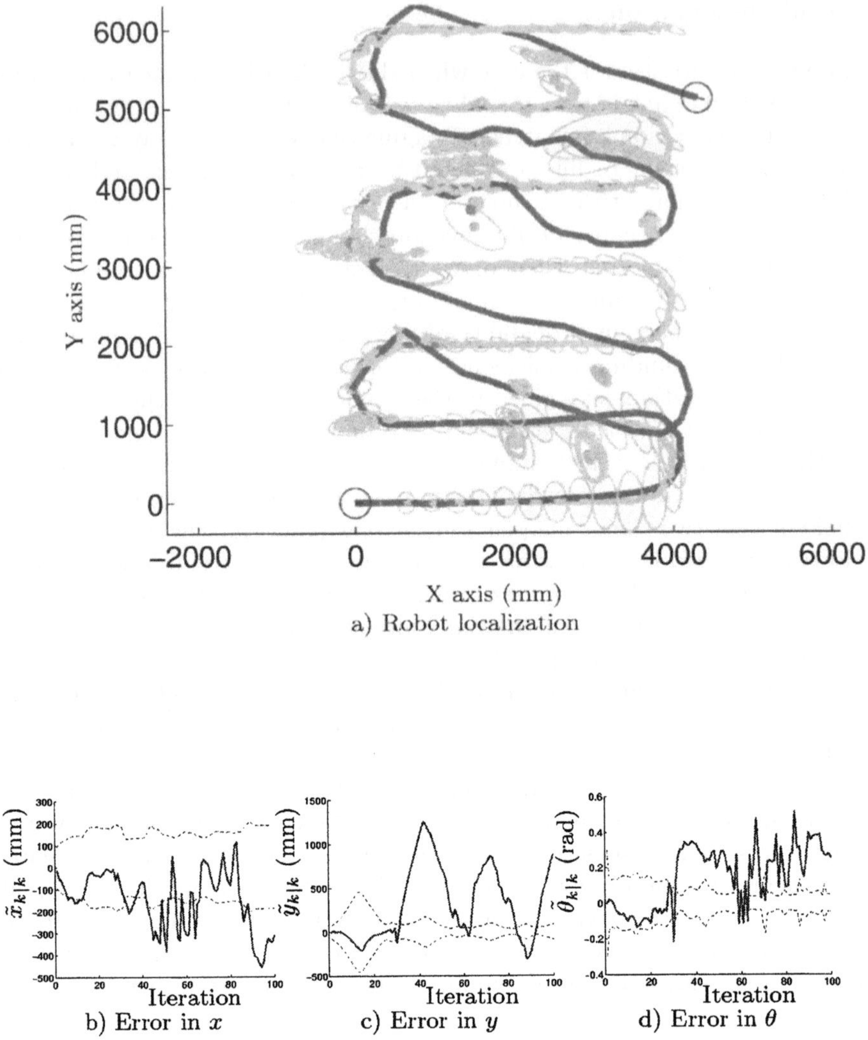

Fig. 1.18. Full-covariance EKF CML for a path with 100 iterations and 10 landmarks, and a sensor with a limited radius of observation of $2m$. Data missassociation occurs within a radius of $1m$ with a 25% probability.

Landmark validation

To correct the limitations of the EKF-SLAM algorithm shown so far, we present now results of the two strategies for the computation of landmark quality herein discussed. First, we will show the results of using the compatibility test to validate landmark observations in terms of their weighted distance to their expected location (in the sense of Mahalanobis d^2). Figure 1.19 shows for our base test case, the improvement in the localization of the mobile robot when the spatial compatibility test from Equation 1.41 (χ^2 goodness of fit test) is performed, with a confidence level of 95%.

The dots in the map plot represent all of the observations that pass the test. Note how landmarks 2 and 9 were misidentified early after they were initialized, and as a consequence their observations do not pass the spatial compatibility test for the rest of the simulation. The estimated robot localization error is represented by the projected ellipses along the robot path, which correspond to 2σ level curves of the robot pose error covariance estimate. The ellipses around landmark locations on the other hand, indicate projected 2σ bounds for the observed landmark covariance estimates.

Also, due to the absence of observations that pass the test during the interval between the 85th and 95th iterations, the vehicle position estimate is revised by dead reckoning only, with the immediate consequence of having the vehicle covariance estimate enlarged at each iteration. Once a landmark passing the test re-enters the robot field of view, the filter kicks in again, with revised vehicle location and error covariance estimates.

Next, we include results of applying the temporal landmark quality test to the above results. Those landmarks whose temporal landmark quality falls below a given threshold are removed from the map. This is, they can no longer be used for localization. Only after observations for those landmarks become robust again, they can be reinitialized. Figure 1.20 shows the improvement of using the exponential decay rule as a function to validate temporal landmark quality, together with individual spatial landmark compatibility tests in SLAM. Furthermore, Figure 1.21 shows the results of using the data association probability as a function to test the temporal landmark quality.

As with Figure 1.19, the dots in the plot represent only the observations that pass the test. Note how observations for landmarks 2 and 9 pass now the test, once they are reinitialized and become robust again. The advantage over the previous simulation is that these landmarks will still be used to build the map, providing more information to the filter (I_T), and consequently converging to a lower localization error estimate. Figure 1.9 shows the evolution of both the exponential decay rule, and the data association probability as landmark quality measures for this test run.

We plot in Figure 1.22 the norm of the robot localization error in the xy plane to show the improvement of using both the temporal landmark quality and spatial landmark compatibility tests, versus using the spatial landmark compatibility test only. The test run is performed on our standard case of the

EKF SLAM algorithm with sequential innovation, a $2m$ limited sensor range, and 25% data missassociation.

New simulations were run for varying conditions, first by increasing the number of landmarks, and then by also incrementing the number of iterations. Figures 1.23 and 1.24 show the results of comparing the use of spatial landmark validation only versus using temporal and spatial landmark validation.

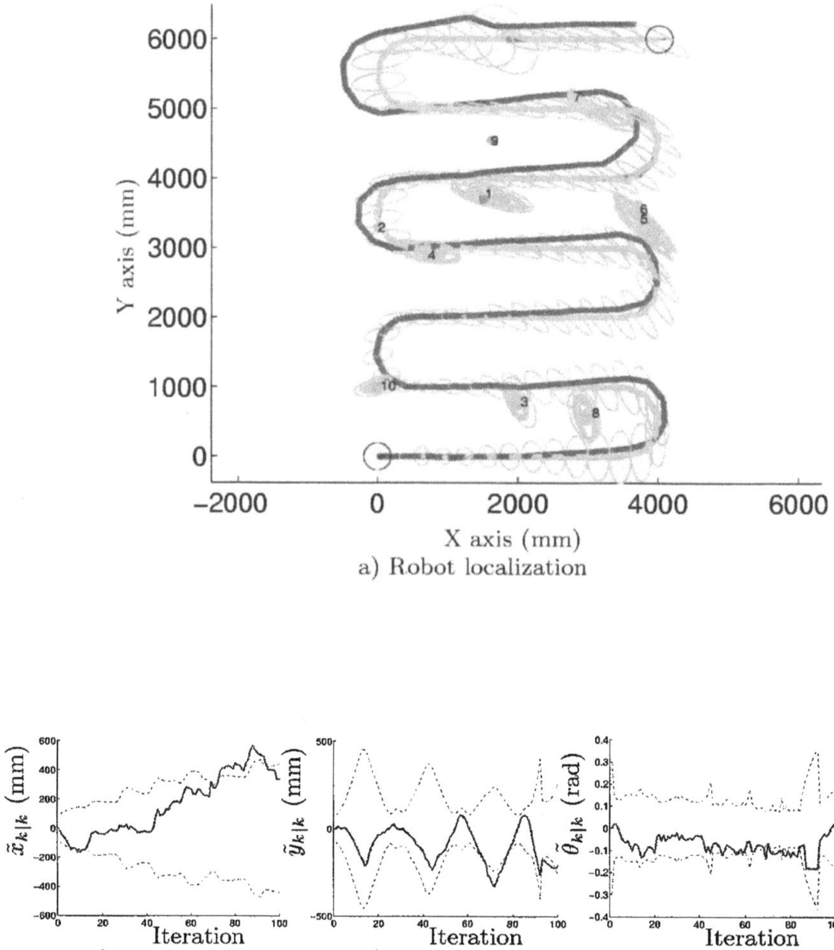

a) Robot localization

b) Error in x

c) Error in y

d) Error in θ

Fig. 1.19. Results of using the individual compatibility test. Full-covariance EKF SLAM for a path with 100 iterations and 10 landmarks, and a sensor with a limited radius of observation of $2m$. Data missassociation occurs within a radius of $1.0m$ with a 25% probability.

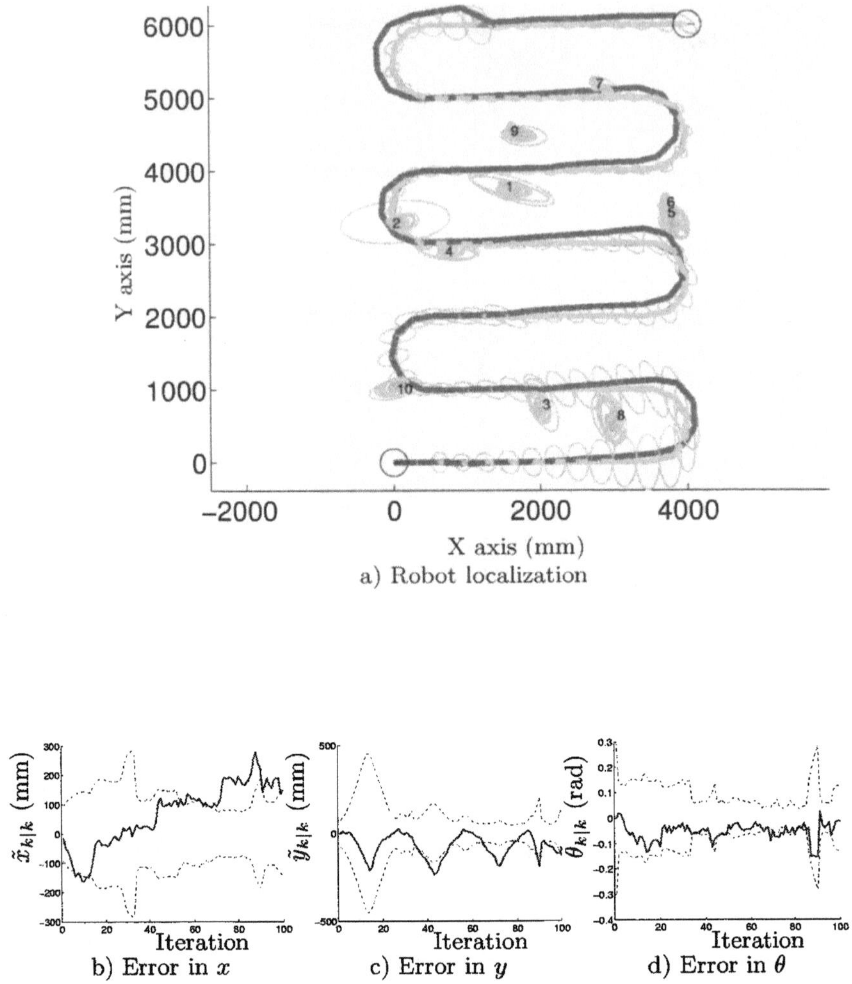

Fig. 1.20. Results of using the individual compatibility test and the exponential decay rule to test temporal landmark quality, $x_{q,LOW} = 0.6590$, $q = 0.66$, $x_{q,HIGH} = 0.8659$. Full-covariance EKF SLAM for a path with 100 iterations and 10 landmarks, and a sensor with a limited radius of observation of $2m$. Data missassociation occurs within a radius of $1.0m$ with a 25% probability.

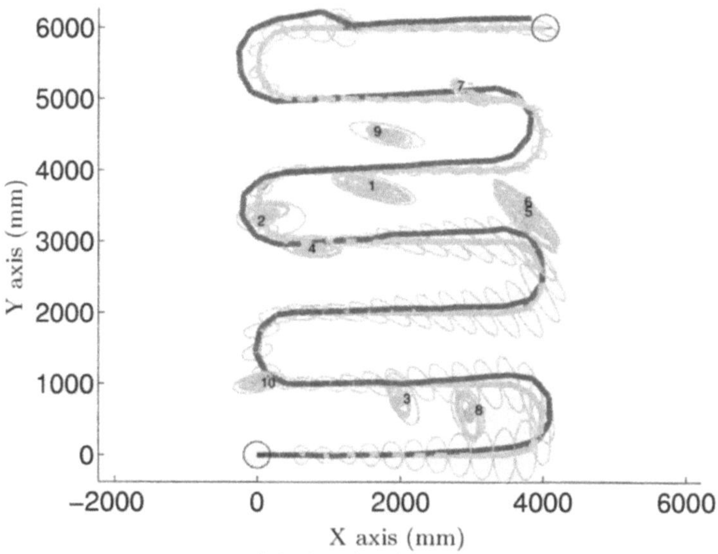

a) Robot localization

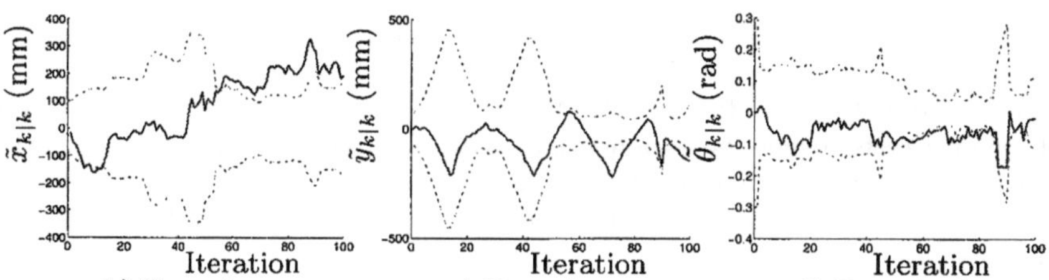

b) Error in x c) Error in y d) Error in θ

Fig. 1.21. Results of using the individual compatibility test and the data association probability as temporal landmark quality test, $x_{q,LOW} = 0$, $q = 0.03$, $x_{q,HIGH} = 1$. Full-covariance EKF SLAM for a path with 100 iterations and 10 landmarks, and a sensor with a limited radius of observation of $2m$. Data missassociation occurs within a radius of $0.5m$ with a 10% probability.

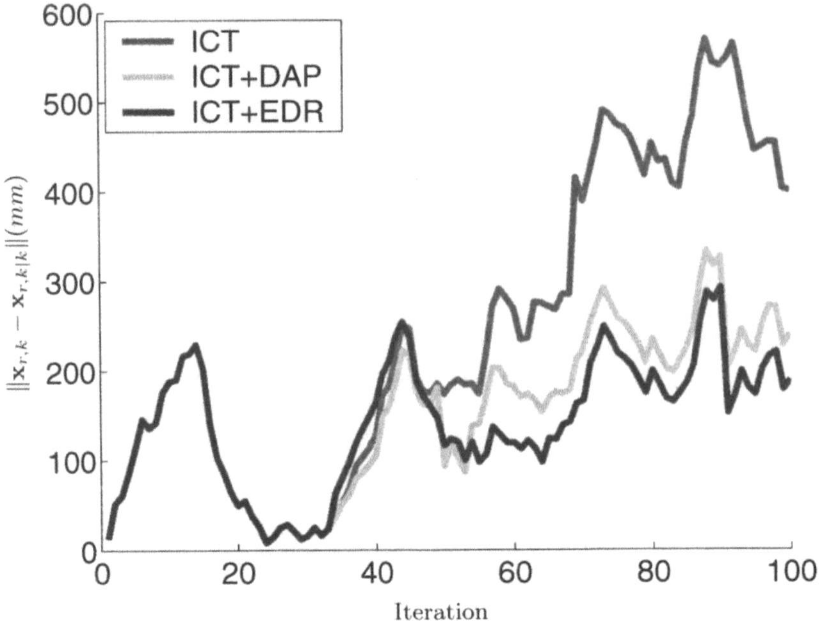

Fig. 1.22. Robot localization error estimate. Comparison between ICT: Individual spatial compatibility test with limited vision range and data association errors (realistic case of SLAM); and the improvements proposed, ICT+DAP: EKF-SLAM-LV with both spatial and temporal landmark quality tests, using the data association probability, and ICT+EDR: EKF-SLAM-LV with both spatial and temporal landmark quality tests, using the exponential decay rule. The simulation is run over a path with 100 iterations and 10 landmarks, and a sensor with a limited radius of observation of $2m$. Data missassociation occurs within a radius of $1.0m$ with a 25% probability.

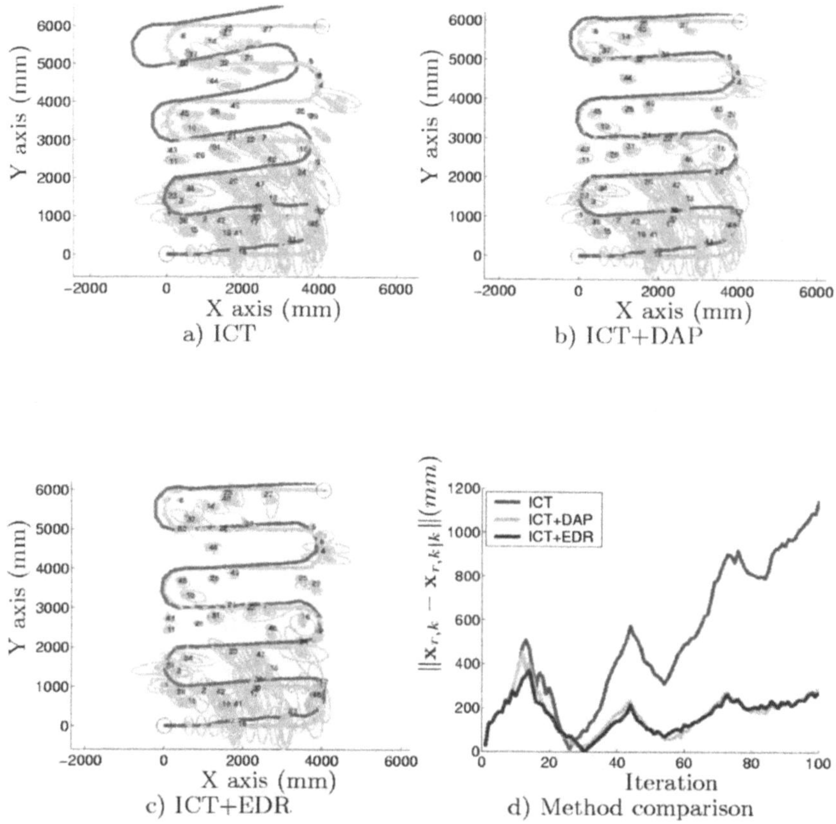

Fig. 1.23. Robot localization error estimate. Comparison between ICT: Individual spatial compatibility test with limited vision range and data association errors (realistic case of SLAM); and the improvements proposed, ICT+DAP: EKF-SLAM-LV with both spatial and temporal landmark quality tests, using the data association probability, and ICT+EDR: EKF-SLAM-LV with both spatial and temporal landmark quality tests, using the exponential decay rule. The simulation is run over a path with 100 iterations and 50 landmarks, and a sensor with a limited radius of observation of $2m$. Data missassociation occurs within a radius of $1.0m$ with a 25% probability.

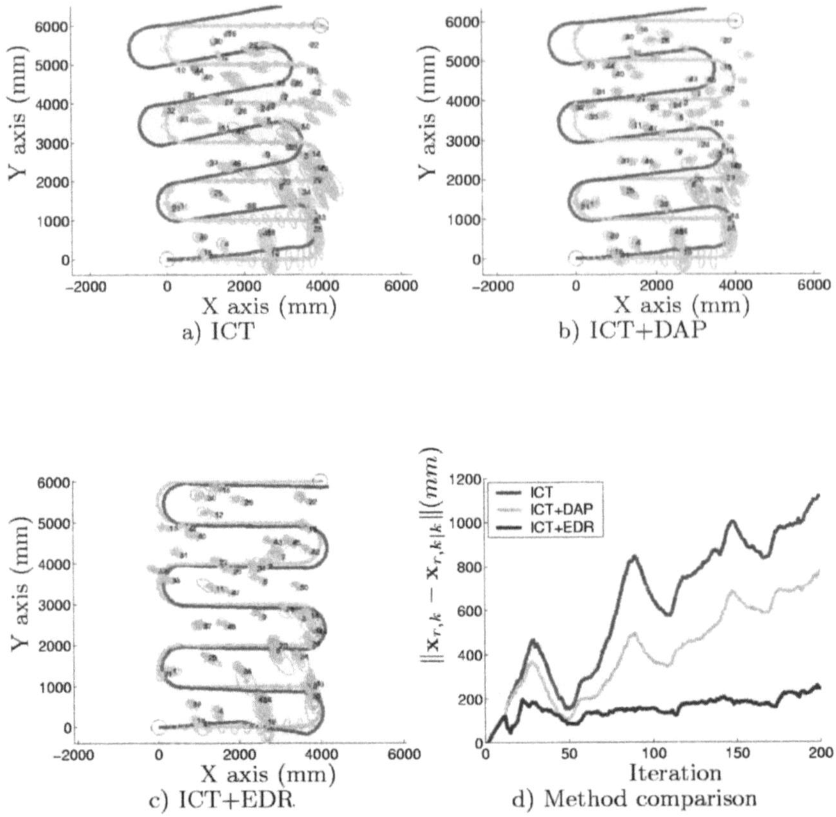

Fig. 1.24. Robot localization error estimate. Comparison between ICT: Individual spatial compatibility test with limited vision range and data association errors (realistic case of SLAM); and the improvements proposed, ICT+DAP: EKF-SLAM-LV with both spatial and temporal landmark quality tests, using the data association probability, and ICT+EDR: EKF-SLAM-LV with both spatial and temporal landmark quality tests, using the exponential decay rule. The simulation is run over a path with 200 iterations and 50 landmarks, and a sensor with a limited radius of observation of $2m$. Data missassociation occurs within a radius of $1.0m$ with a 25% probability.

1.5 Divergence

There is unfortunately a tradeoff in using the Extended Kalman Filter in SLAM. Even with perfect data association, for nonlinear vehicle models, the algorithm diverges in the long run. The divergence has been attributed to the linearization in Equations 1.6 and 1.7. An explanation of this behavior is by Julier and Uhlmann [53], where they show a counter example to the EKF-SLAM algorithm that did not converge to a fully correlated map in their simulations.

Their vehicle model was similar to our planar mobile robot from Section 1.2; a vehicle with three state variables to indicate its pose, two Cartesian coordinates for position, and an orientation angle. In their simulations they showed how such a vehicle could not reconstruct its location from continuous observations to one point landmark; even, when no motion commands were conveyed to the robot, and when a noise free plant model was assumed.

They suggest that the divergence is due to the linearization of the system needed for the implementation of the EKF, arguing that the no-motion condition for an error-free model is equivalent to having a zero component in the vehicle part of the Kalman gain. Such assumption leads to the equality

$$\mathbf{H}_r^{(i)}\mathbf{F}\mathbf{x}_r = -\mathbf{H}_f^{(i)}\mathbf{x}_f^{(i)} \tag{1.72}$$

Furthermore, they show how Equation 1.72 holds only for the linear case of SLAM, and breaks down after linearization in the EKF version of the algorithm, and state that "being this a structural condition, no tuning procedure (inflation of plant and measurement noise covariances) would circumvent the problem". For this reason they push for nonlinear approaches to KF. To palliate the effects of nonlinearities we analyze in Chapter 4, one such technique to propagate means and covariances without the need for computing model Jacobians.

Moreover, in the next Chapter we analyze how the localization error varies with respect to the number of landmarks used. Our study is from the point of view of control systems theory. We believe that the amount of divergence of the algorithm may also be factored by the partial observability of the state vector, and that by keeping the state error covariance as close to the lower bound $\mathbf{P}_{r,0|0}$ as possible, one can reduce the divergence effects. This can only be done if we add more landmarks to the map, or ultimately, as we propose, by anchoring the map to a known world reference.

1.6 Bibliographical Notes

Mobile robot localization has been addressed from different perspectives, divided mainly in the following groups.

Correlation methods that match sensor signals against previously stored maps. Systems that match visual data using correlations for self-localization include the ones by Drocourt *et al.* [33], Hashima *et al.* [47], Mallet and Lacroix [66], Talluri and Aggarwal [82], or Volpe *et al.* [90].

Estimation theoretic approaches that predict and refine the robot position from current and previous sensor readings, past position estimates, and motion commands, as well as uncertainty models of sensors and motion. Such methods are typically robust for local localization, provided that the initial estimate of the robot is sufficiently accurate, and that correct stochastic models for the robot dynamics and the sensors are available. An example where vertical lines are extracted from images and matched against a previously stored map of the environment taking into account the stochastic nature of sensor data is presented in the work by Ohya *et al.* [72]. In this case, robust self-localization is attained by means of Kalman filtering. Recent contributions in the use of Kalman filtering techniques for self-localization include those of Anousaki and Kyriakopoulos [15], Deans and Hebert [26], Duckett and Nehmzow [34], Kosaka and Kak [60], Lee and Recce [62], and our own contributions deriving from this work [6, 7, 8, 9, 10, 11, 12, 13, 14, 88, 89].

Markov localization techniques are an extension to the above cited method, in which a probabilistic framework is used to maintain a position probability density over the whole set of robot poses. These techniques are better suited to solve the global localization problem, at the expense of stronger assumptions about the nature of the environment than with the Kalman filter. A recent example that illustrates the use of Markov models for robot localization is by Fox *et al.* [38]. A thorough description of Markov localization methods can be found in the work by Thrun [84].

Finally, a mirage of other algorithms have been proposed in attempts to solve the robot localization problem. The ones that are less related to the work presented in this thesis vary from fuzzy-based algorithms [28, 76, 77], to the *minmax risk fixed-size confidence set estimate* [67], designed as an optimal set of decision rules in the minmax sense for dealing with the uncertainties in sensor measurements; a topological approach that makes use of a generalized Voronoi graph and graph matching techniques [22]; or a set-theoretic approach, in which the estimate of the position of the robot and the landmarks is expressed in terms of bounded sets, making use of set membership estimation theory.

The main drawback of most of these approaches is however, that they have not been designed with continuously changing environments in mind. All of these methods treat moving or temporary objects as noisy data or measurement uncertainty. It is only with Kalman filter based methods and Markov localization techniques that by using statistical approaches we can cope with noisy data. Fuzzy based algorithms do the same by estimating qualitative models for sensor and robot uncertainty. We believe that by incorporating time-varying properties to our environment model, we are able to better cope with the dynamics inherent in typical mobile robotics environments.

The study of stochastic models for simultaneous localization and map building in mobile robotics has been addressed by several research groups for over fifteen years. Within the KF approach to SLAM, seminal work by Smith and Cheeseman [79] suggested that as successive landmark observations take place, the correlation between the estimates of the location of such landmarks in a map grows continuously. This observation was ratified recently by Dissanayake *et al.* [31] with a proof showing that the estimated map converges monotonically to a relative map with zero uncertainty. They also showed how the absolute accuracy of the map reaches a lower bound defined only by the initial vehicle uncertainty. With respect to covariance initialization, we would like to express our gratitude to P. Newman for making it clear to us. Thanks to that we were able to reproduce in our simulations the three asymptotic convergence properties of SLAM described in Section 1.1.

In spite of these fundamental convergence properties of the KF approach to SLAM, there exist some limitations that still hinder full development of SLAM applications. The three most criticized of these limitations are the time and space complexity of the algorithm, the restriction to unimodal zero mean white Gaussian models of uncertainty, and the data association and landmark quality problems. We go even further in our analysis of the SLAM problem, and show in the next chapter one more fundamental limitation to the algorithm, that full reconstruction of the map state vector is not possible, regardless of the vehicle model chosen, and that the expected error in state estimation is proportional to the number of landmarks used.

The explicit solution for the monobot SLAM problem presented by Gibbens *et al.* [41] shed some light on the relation between the total number of landmarks (in the form of the total Fisher information I_T), and the asymptotic values for $\mathbf{P}$. In the next chapter we treat this topic from a different perspective, with the principles of controllability and observability of linear systems in mind.

On a side issue, special interest has recently been put on the search for methods to reduce the time and space complexity of the SLAM algorithm, at the expense of a sub-optimality in the solution; by pruning the map from those landmarks that surpass a certain degree of spatial correlation [30], or by local updating of the map exploiting the advantages of sequential innovation in KF [43]. One requisite for sequential innovation is that observations must be uncorrelated, and they present a suboptimal decorrelation algorithm for that purpose.

Julier *et al,* also present a suboptimal solution to SLAM with sequential innovation in KF, reducing the time and space complexity of the algorithm from $O(n^2)$ to $O(n)$ [49, 54]. Instead of decorrelating observations, they do away with cross correlations between observations and state estimates, and update the state covariance estimate with a weighting factor that penalizes the lack of information about correlations.

In the search for robust models of uncertainty in SLAM, a variety of routes have been explored. Durrant-Whyte *et al.* presented an extension to the typical KF-SLAM algorithm with the use of sums of Gaussians (SOGs), to approximate more general probability distributions for the modeling of unstructured sub-sea terrain [36]. Tardós *et al.* [83] build maps of indoor environments from sonar data, suing the Hough transform as a tool to perceptual grouping of sonar returns, and by joining local maps into a global representation.

There is no general consensus on what constitutes a good measure for landmark quality. Intuition suggests measures proportional to the temporal dispersion and inversely proportional to the spatial dispersion of landmarks. We have no knowledge of previous attempts at formulating temporal landmark quality models. There resides precisely one of the most relevant contributions of this thesis. In an attempt to unify landmark quality and data association uncertainty within the structure of the estimation-theoretic approach to SLAM, we proposed in this Chapter, an augmented map model that incorporates temporal landmark temporal tests to validate the quality of observations.

Dissanayake *et al.* address the issue of landmark robustness as an implementation detail only [31], suggesting a quality measure based on the probability density function of the observations associated to any given landmark, and disregarding the temporal dispersion of such observations. A different, more simple measure of landmark quality is proposed [30], as the trace of the error covariance submatrix $\mathbf{P}_{f_i}$.

To aid in data association, Castellanos *et al.* suggested in [21], a χ^2 test to evaluate scene-to-map landmark match hypotheses based on the squared Mahalanobis distance between map landmarks and observations. Here, the uncertainty in the location of a landmark in the map plays the role of quality assessment. More recently, Neira and Tardós [70] presented a constrained search algorithm for scene to map hypothesis formation, in which the use of the joint compatibility test is argued to supersede the classical nearest neighbor compatibility test typically used for independent matching of sensors and measurements.

Leonard *et al.* [64, 65] have realized that spatial dispersion is not sufficient for landmark quality assessment, and that temporal information is needed as well. They have opted for a delayed decision approach to data association, in which by adding previous robot states to the state vector, they can include in the Kalman filter covariance matrix correlations pertaining estimations from previous vantage points. In this way, they can add non-invariant features such as range-only measurements to aid in data association. A serious disadvantage of this technique is that the map grows linearly in time, increasing the computational burden of the SLAM algorithm, even when no new features are added to the map. Another serious disadvantage is the divergence of the fundamental motivation of using a Kalman filter to maintain a parametric representation of the history of the uncertainty of robot and landmark localization by means of an estimated full-covariance matrix.

Davison *et al.* have also studied the spatial dispersion of landmark observations as a measure of landmark quality, and suggest that when one has to choose among several observations to update, the one with the largest hyper-ellipsoid volume of dispersion contributes more to the overall reduction of the map error covariance. Similarly, they argue that by diagonalizing the landmark covariance in measurement space $\mathbf{H}_i\mathbf{P}\mathbf{H}_i^\top$, they are able to bound the search for scene-to-model landmark matches. These observations led them to active vision heuristics for gaze control when building maps with SLAM [24, 25]. Another suggestion is the use of robust methods aimed at eliminating observation outliers, such as RANSAC [37], in order to diminish the effect of measurement and data association errors.

General references to linear systems theory include the books by Kailath [56], and DeCarlo [27]. For the study of the Pearson χ^2 test and other topics of statistics the reader is referred to the book by Fukunaga [39]. Linear Gaussian models in general are treated extensively in the survey by Roweis and Ghahramani [75], and in the book by Duda *et al.* [35].

1.7 Concluding Remarks

This chapter presents a revision of the traditional full-correlation EKF SLAM algorithm for mobile robot localization and map building. We extend the traditional algorithm by adding temporal landmark quality measures, and a temporal landmark quality test to validate the history of data association. These quality measures permit the maintenance of the map by the elimination of inconsistent observations. The removal of weak landmarks from the state vector and state covariance matrix does not violate the convergence properties of SLAM. Special attention has been paid in the selection of the temporal landmark quality models, to guarantee that the uncertainty in the map estimates still reduces monotonically. The proposed solution contributes in simplifying the data association problem in SLAM.

One drawback that might limit the use of the newly introduced temporal landmark quality test in computer vision based SLAM is the concept of visibility. Temporal landmark quality can only be revised for those landmarks that are visible to the vehicle. And, depending on the sensor and landmark models used, a visibility condition might be rather difficult to assert. Occlusions of walls inferred from laser data are easy to compute; however to test if a visual landmarks should be present or not in a scene, is rather complicated.

2

Marginal Filter Stability

Two fundamental aspects in the use of the Kalman Filter for simultaneous localization and mapping are the issues of observability and controllability. Unfortunately, in SLAM, the state space constructed by appending the robot pose and the landmark locations is fully correlated; a situation that hinders full observability. Moreover, the modeling of map states as static landmarks yields a partially controllable state vector. The identification of these problems, and the steps taken to palliate them, constitute one of the main topics of this monograph. The bulk of which is covered in this chapter.

In Chapter 1 we dedicated a few paragraphs to the formulation of the EKF-SLAM equations for a simple linear 1-D robot we called, the monobot. The importance of the formulation of a naive case study such as the monobot becomes evident in this Chapter, where we study the behavior of the estimation-theoretic approach to SLAM from a systems point of view. We need this simple linear model to show the effects of partial observability and partial controllability, with respect to the number of landmarks used, even when data association is perfect. That is, when all landmarks are correctly identified at each and every iteration.

In Section 2.1 we analyze the steady state behavior of the error state covariance for the monobot SLAM, and show by example, that the steady state of the filter will always depend on the initial noise parameters. The effect is known as filter instability, and is in general an undesirable feature in state estimation.

In Section 2.2 we derive an expression for the total Fisher information in SLAM, and show how full correlation prohibits the use of the Cramer Rao lower bound for the vehicle and map covariance. The analysis yields a closed form solution for the monobot Fisher information matrix that shows explicitly, the unobservable directions of the map state in SLAM.

Filter instability, and the singularity of the Fisher information matrix, are both consequences of having partial observability and controllability. Section 2.4 is devoted to the computation of general expressions for the bases of the controllable and observable subspaces in SLAM. These expressions are later simplified in Sections 2.5 and 2.6 for the monobot and planar vehicle models, respectively. In the end, we show that the angle between these two subspaces is determined only by the total number of landmarks in the map. The result is that as the number of landmarks is incremented, the vehicle pose states get closer to being reconstructible.

The observability condition guarantees a *steady flow* of the information about each state component, and prevents the uncertainty (error state covariance) from becoming unbounded [18]. In Section 2.7 we show how partial observability in SLAM can be avoided by adding a fixed external sensor to the state model, or equivalently, by setting a fixed landmark in the environment to serve as global localization reference. Full observability yields the existence of a (not necessarily unique) steady state *psd* solution for the error covariance matrix **P**.

The controllability condition allows the process noise to enter into each state component and prevents the covariance matrix from becoming singular [18]. In Section 2.8 we show how having a semi-definite covariance matrix reflects on the filter's belief that it has perfectly accurate estimates of some state components, for which the Kalman gain evaluates to zero, and the innovations are not considered in the revision of the state estimate. To palliate the effects of partial controllability, we propose the addition of artificial positive definite process noise associated to the landmark states.

2.1 Steady State Behavior of EKF-SLAM

We start the discussion with a pictorial representation of the behavior of the EKF-SLAM algorithm. Consider a monobot with initial vehicle localization variance $\sigma^2_{r,0|0} = 1$, and various levels of motion and sensor noise variances, denoted by σ^2_v and σ^2_w, respectively. For the sake of simplicity, imagine a robot with constant motion uncertainty, regardless of the size of the input command, and a sensor with constant observation uncertainty, regardless of the distance to the landmark being observed. Later, we will take into account more realistic models of uncertainty, and analyze the advantages of having a measurement covariance matrix that can be computed online.

The final vehicle and landmark variances ($\sigma^2_{r,k|k}$, $\sigma^2_{f,k|k}$, $k \to \infty$) depend only on the initial parameters $\sigma^2_{r,0|0}$, σ^2_v, σ^2_w, and on the total number of landmarks n. The evolution of the error covariance matrix is independent of the state input, and measurements throughout the run of the algorithm. Meaning that, for a monobot with perfect data association and constant motion and sensor uncertainty, the computation of the Kalman gain could even be performed off-line. That is, the asymptotic (steady state) behavior of the filter, and its rate of convergence are always the same, regardless of the actual motions and measurements.

Figure 2.1 shows the steady state vehicle and landmark variances of the EKF-SLAM algorithm applied to a monobot when observations of 1, 2, 3,

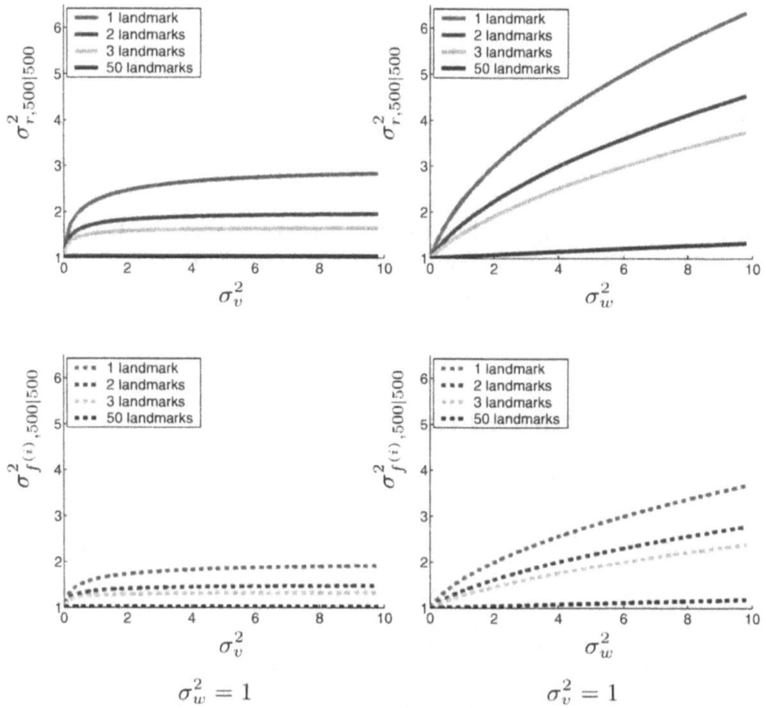

Fig. 2.1. Final vehicle and landmark localization variances after 500 iterations of CML for a monobot with initial localization variance $\sigma^2_{r,0|0} = 1$, and various levels of plant and sensor noise.

and 50 landmarks are available. The Figure plots the influence of each of the noise parameters σ_v^2 and σ_w^2 with respect to the final vehicle and landmark uncertainty.

On the one hand, the final vehicle localization variance depends less on the vehicle plant noise variance σ_v^2 than on the measurement noise variance σ_w^2. On the other hand, as the number of landmarks grows, we see a considerable decrease of the steady state value for all the entries in $\mathbf{P}$. The landmark-to-vehicle and landmark-to-landmark cross correlation terms ($\rho_{rf^{(i)}}\sigma_r\sigma_{f^{(i)}}$ and $\rho_{f^{(i)}f^{(j)}}\sigma_{f^{(i)}}\sigma_{f^{(j)}}$) although not shown on the plots, they all converge to the same values as the landmark variance $\sigma^2_{f^{(i)}}$. Meaning that, as k tends to infinity, the map becomes fully correlated, i.e., $\rho_{f^{(i)}f^{(j)}} = 1$.

The two observations: a) that rate of convergence of $\mathbf{P}$ is fixed, and b) that the precision of the map (in terms of the asymptotic value of the error covariance) is independent of the plant variance σ_v^2; were first described by Gibbens *et al.* [41], in which a closed form solution for the computation of Equation 2.1 for the continuous time domain monobot was presented.

The steady state covariance matrix is given by the solution of the Riccati equation

$$\mathbf{P} = \mathbf{F}(\mathbf{P} - \mathbf{P}\mathbf{H}^\top(\mathbf{H}\mathbf{P}\mathbf{H}^\top + \mathbf{R})^{-1}\mathbf{H}\mathbf{P})\mathbf{F}^\top + \mathbf{Q} \tag{2.1}$$

and for the linear case it is only a function of $\mathbf{P}_{rr,0|0}$, $\mathbf{Q}$, $\mathbf{R}$, and n. Note however that, for the nonlinear case, the computation of the Jacobians $\mathbf{F}$ and $\mathbf{H}$ will in general also depend on the steady state value of $\mathbf{x}$.

For a linear robot with perfect data association and constant vehicle and sensor variances, the computation of $\mathbf{K} = \mathbf{P}\mathbf{H}^\top(\mathbf{H}\mathbf{P}\mathbf{H}^\top + \mathbf{R})^{-1}$ could be performed off-line. That is, the asymptotic behavior of $\mathbf{P}$ and its rate of convergence are always the same, regardless of the actual motions and measurements.

2.2 Total Fisher Information

Under the Gaussian assumption for the vehicle and sensor noises, the Kalman filter is the optimal minimum mean square error estimator. And, as pointed out in [18], minimizing the least squares criteria from Equation A.17

$$E[\tilde{\mathbf{x}}_{k|k}\tilde{\mathbf{x}}_{k|k}^\top]$$

is equivalent to the maximization of a likelihood function $\Lambda(\mathbf{x})$ given the set of observations Z^k; that is, the maximization of the joint probability density function of the entire history of observations, conditioned on the state $\mathbf{x}$

$$\begin{aligned}\Lambda(\mathbf{x}) &= p(Z^k|\mathbf{x}) \\ &= p(\mathbf{z}_k, Z^{k-1}|\mathbf{x}) \\ &= p(\mathbf{z}_k|\mathbf{x}, Z^{k-1})p(Z^{k-1}|\mathbf{x}) \\ &= \prod_{i=1}^{k} p(\mathbf{z}_i|\mathbf{x}, Z^{i-1})\end{aligned} \tag{2.2}$$

Given that the above pdfs are Gaussian, and that $E[\mathbf{z}_i] = \mathbf{H}\mathbf{x}_{i|i-1}$, the pdf for each measurement in SLAM is

$$\begin{aligned}p(\mathbf{z}_i|\mathbf{x}, Z^{i-1}) &= N(\tilde{\mathbf{z}}_{i|i-1}; \mathbf{0}, \mathbf{S}_i) \\ &= (2\pi)^{-\frac{dn}{2}}|\mathbf{S}_i|^{-\frac{1}{2}}e^{-\frac{1}{2}(\tilde{\mathbf{z}}_{i|i-1}^\top\mathbf{S}_i^{-1}\tilde{\mathbf{z}}_{i|i-1})}\end{aligned} \tag{2.3}$$

That is, the joint pdf of the sequence of measurements Z^k is equal to the product of the marginal pdfs of the corresponding innovations.

In practice however, it is more convenient to consider the log likelihood function

$$\begin{aligned}\ln \Lambda(\mathbf{x}) &= \sum_{i=1}^{k} \ln p(\mathbf{z}_i|\mathbf{x}, Z^{i-1}) \\ &= -\frac{1}{2}\sum_{i=1}^{k} \tilde{\mathbf{z}}_{i|i-1}^\top\mathbf{S}_i^{-1}\tilde{\mathbf{z}}_{i|i-1} + \sum_{i=1}^{k} \ln|2\pi\mathbf{S}_i|\end{aligned} \tag{2.4}$$

An intuitive interpretation of the maximum of the log-likelihood in Equation 2.4 is that the best estimate for the state $\mathbf{x}$, in the least squares sense, is the one that makes the sum of the entire set of Mahalanobis distances $\sum_{i=1}^{k} \tilde{\mathbf{z}}_{i|i-1}^{\top} \mathbf{S}_i^{-1} \tilde{\mathbf{z}}_{i|i-1}$ as small as possible. A measure that is consistent to the spatial compatibility test described in Section 1.1.

The maximum of $\ln \Lambda(\mathbf{x})$ is at the value of the state $\mathbf{x}$ that most likely gave rise to the observed data Z^k, and is obtained by setting the derivative of Equation 2.4 with respect to $\mathbf{x}$ equal to zero, which gives

$$\nabla_{\mathbf{x}} \ln \Lambda(\mathbf{x}) = \sum_{i=1}^{k} \mathbf{H}^{\top} \mathbf{S}_i^{-1} \tilde{\mathbf{z}}_{i|i-1} \tag{2.5}$$

The Fisher information matrix, a quantification of the maximum existing information in the observations about the state $\mathbf{x}$ is defined (in [17] and [18]) as the expectation on the dyad of the gradient of Equation 2.4

$$\mathbf{J} = E[(\nabla_{\mathbf{x}} \ln \Lambda(\mathbf{x}))(\nabla_{\mathbf{x}} \ln \Lambda(\mathbf{x}))^{\top}] \tag{2.6}$$

Taking the expectation on the innovation error $E[\tilde{\mathbf{z}}_{i|i-1} \tilde{\mathbf{z}}_{i|i-1}^{\top}] = \mathbf{S}_i$ in the above formula gives the sum

$$\mathbf{J} = \sum_{i=1}^{k} \mathbf{H}^{\top} (\mathbf{H}\mathbf{P}\mathbf{H}^{\top} + \mathbf{R})^{-1} \mathbf{H} \tag{2.7}$$

In the linear case, this expression for the total Fisher information is only a function of $\mathbf{P}_{rr,0|0}$, $\mathbf{Q}$, and $\mathbf{R}$. If on the other hand, the EKF has been used, the Jacobian $\mathbf{H}$ in Equation 2.6 should be evaluated at the true value of the states $\mathbf{x}_0, \ldots \mathbf{x}_k$. Since these are not available, an approximation is obtained at the estimates $\mathbf{x}_{i|i-1}$. The pre and post multiplying Jacobian $\mathbf{H}$ in Equation 2.7 is, in this context, also known as the *sensitivity matrix* [32].

A necessary condition for the estimator (the Kalman filter) to be consistent in the least squares sense is that there must be an increasing amount of information, about the state $\mathbf{x}$ in the measurements. That is, as $k \to \infty$, the Fisher information has to tend to infinity.

Figure 2.2 shows this for the monobot with constant parameters $\sigma_{r,0|0}^2 = \sigma_v^2 = \sigma_w^2 = 1$, and various sizes for the observation vector. Notice how, as the total number of landmarks grows, the total Fisher information also grows, directly relating the number of landmarks to the amount of information available for state estimation in SLAM.

Solving for the k-th sum term in $\mathbf{J}$ for the monobot yields the expression

$$\mathbf{J}_k = \begin{bmatrix} \sum\sum \varsigma_{ij} & -\boldsymbol{\varsigma} \\ -\boldsymbol{\varsigma}^{\top} & \mathbf{S}_k^{-1} \end{bmatrix} \tag{2.8}$$

with ς_{ij} the ij-th entry in $\mathbf{S}_k^{-1}$, and $\boldsymbol{\varsigma} = [\sum \varsigma_{1i}, \ldots, \sum \varsigma_{ni}]$.

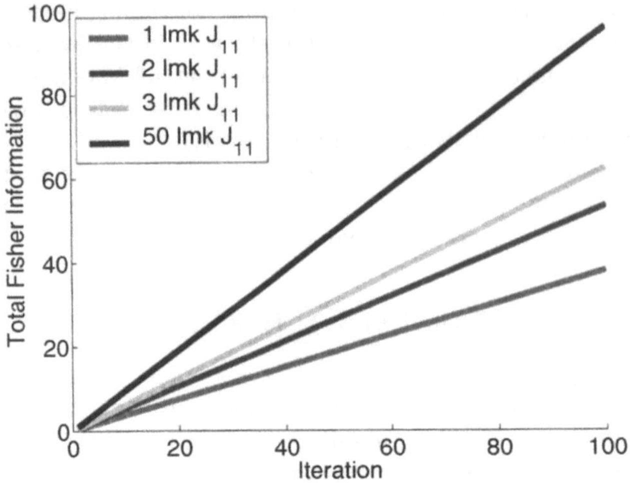

Fig. 2.2. First entry in the total Fisher information matrix for a monobot with variance parameters $\sigma^2_{r,0|0} = \sigma^2_v = \sigma^2_w = 1$, and various sizes for the measurement vector.

Citing Bar-Shalom *et al.* [18]: "*A lower bound on the minimum achievable covariance in state estimation is given by the posterior Cramer Rao lower bound*"[1]

$$E[\tilde{\mathbf{x}}_{k|k}\tilde{\mathbf{x}}^\top_{k|k}] \geq \mathbf{J}^{-1} \tag{2.9}$$

Unfortunately, it can be easily shown, at least for the monobot case, that the first row (or column) of $\mathbf{J}$ is equivalent to the sum of the rest of the rows (or columns), producing a singular total Fisher information matrix. In SLAM the Cramer Rao lower bound cannot be evaluated. SLAM is unobservable.

Citing once more Bar-Shalom *et al.*: "*if the Fisher information matrix is not invertible, then the lower bound from Equation 2.9 will not exist, actually it will have one or more infinite eigenvalues* (one in the case of SLAM), *which means total uncertainty in a subspace of the state space, that is, the information is insufficient for the estimation problem at hand.*"

This is a consequence of the form of the Jacobian $\mathbf{H}$, i.e, of the full correlation in SLAM. Zero eigenvalues of $\mathbf{H}^\top\mathbf{S}^{-1}\mathbf{H}$ are an indicator of partial observability, and the corresponding vectors give the unobservable directions in state space.

So for example, for a one-landmark monobot, the innovation covariance is the scalar $s = \sigma^2_r - 2\rho_{rf}\sigma_r\sigma_f + \sigma^2_f + \sigma^2_w$, and since $\mathbf{H} = [-1, 1]$, the Fisher information matrix in Equation 2.8 evaluates to

[1] See Appendix B for our interpretation of matrix inequalities.

$$\mathbf{J} = \begin{bmatrix} 1 & -1 \\ -1 & 1 \end{bmatrix} \sum_{i=1}^{k} \frac{1}{s_i} \tag{2.10}$$

The unobservable direction of the state space is the eigenvector associated to the null eigenvalue of $\mathbf{J}$, we denote it for now $\mathbf{E}_{\mathrm{Ker}\mathcal{O}}$ (the name will be clear soon), and evaluates to

$$\mathbf{E}_{\mathrm{Ker}\mathbf{R}} = \begin{pmatrix} 1 \\ 1 \end{pmatrix} \tag{2.11}$$

2.3 Convergence

Substituting the linearized version of (1.4) in (1.19), we may rewrite the KF in the one-step ahead prediction form

$$\mathbf{x}_{k+1|k} = (\mathbf{F} - \mathbf{KH})\,\mathbf{x}_{k|k-1} + \mathbf{Kz}_k \tag{2.12}$$

and with the appropriate substitutions, using (1.16) and (2.12), the corresponding prediction error dynamics becomes

$$\tilde{\mathbf{x}}_{k+1|k} = (\mathbf{F} - \mathbf{KH})\tilde{\mathbf{x}}_{k|k-1} + \mathbf{Gv}_k - \mathbf{Kw}_k \tag{2.13}$$

In general, only for a stable matrix $\mathbf{F} - \mathbf{KH}$, the estimation error will converge to a zero mean steady state value. However, in SLAM, $\mathbf{F} - \mathbf{KH}$ is marginally stable, thus the steady state error estimate is bounded to a constant value, subject to the filter initial conditions. To show $\mathbf{F}-\mathbf{KH}$ marginally stable, consider a one landmark monobot, i.e., $\mathbf{F} = \mathbf{I}$, $\mathbf{G} = [1\;0]^\top$, and $\mathbf{H} = [-1\;1]$. For any value of

$$\mathbf{P} = \begin{bmatrix} \sigma_r^2 & \rho\sigma_r\sigma_f \\ \rho\sigma_r\sigma_f & \sigma_f^2 \end{bmatrix} \tag{2.14}$$

the Kalman gain, computed with (1.20), is

$$\mathbf{K} = \frac{1}{s} \begin{bmatrix} -\sigma_v^2 \\ \sigma_w^2 \end{bmatrix} \tag{2.15}$$

where

$$s = \sigma_r^2 + \sigma_f^2 - 2\rho\sigma_r\sigma_f + \sigma_w^2 \tag{2.16}$$

is the innovation variance. Consequently,

$$\mathbf{F} - \mathbf{KH} = \frac{1}{s} \begin{bmatrix} -\sigma_v^2 + s & \sigma_v^2 \\ \sigma_w^2 & -\sigma_w^2 + s \end{bmatrix} \tag{2.17}$$

with eigenvalues

$$\left\{ \begin{matrix} 1, \\ \frac{1}{s}\left(s - \sigma_v^2 - \sigma_w^2\right) \end{matrix} \right\} \text{ and } s \neq 0$$

One of the eigenvalues being on the unitary circle yields marginal stability, i.e., constant bounded non-zero mean error state estimate convergence.

2.4 Observable and Controllable Subspaces

The solution to the Riccati Equation 2.1 converges to a finite steady state covariance if the pair $\{\mathbf{F}, \mathbf{H}\}$ is completely observable. If in addition, the pair $\{\mathbf{F}, \mathbf{G}\}$ is completely controllable, then the steady state covariance is a unique positive-definite matrix, independent of the initial covariance $\mathbf{P}_{0|0}$.

Partial observability in SLAM means that the information entering the system in the form of innovations is fully correlated, and that there is no guarantee that the state error covariance is bounded. In other words, given the singularity of the Fisher information matrix $\mathbf{J}$, the information extracted from the innovations $\tilde{\mathbf{z}}$ is insufficient for the entire reconstruction of the state estimate $\mathbf{x}$.

To see what part of the state space is compromised by full correlation, we now develop closed form expressions for the bases of the observability and controllability subspaces in SLAM and relate them to the total number of landmarks used.

The linearized EKF state model can be rewritten in terms of the error states with

$$\tilde{\mathbf{x}}_{k+1|k} = \mathbf{F}\tilde{\mathbf{x}}_{k|k} + \mathbf{G}\mathbf{v}_k \tag{2.18}$$

$$\tilde{\mathbf{z}}_{k+1|k} = \mathbf{H}\tilde{\mathbf{x}}_{k+1|k} + \mathbf{w}_{k+1} \tag{2.19}$$

Combining the plant and measurement noises into one large input vector, the above error driven system is equivalent to the augmented model

$$\tilde{\mathbf{x}}_{k+1|k} = \mathbf{F}\tilde{\mathbf{x}}_{k|k} + \begin{bmatrix}\mathbf{G}\ \mathbf{0}\end{bmatrix}\begin{bmatrix}\mathbf{v}_k \\ \mathbf{w}_{k+1}\end{bmatrix} \tag{2.20}$$

$$\tilde{\mathbf{z}}_{k+1|k} = \mathbf{H}\mathbf{F}\tilde{\mathbf{x}}_{k|k} + \begin{bmatrix}\mathbf{H}\mathbf{G}\ \mathbf{I}\end{bmatrix}\begin{bmatrix}\mathbf{v}_k \\ \mathbf{w}_{k+1}\end{bmatrix} \tag{2.21}$$

and the controllability matrix for such a plant is

$$\mathcal{C} = \begin{bmatrix}\mathbf{G}\ \mathbf{0} | \mathbf{F}\mathbf{G}\ \mathbf{0} | \ldots | \mathbf{F}^{m+dn-1}\mathbf{G}\ \mathbf{0}\end{bmatrix} \tag{2.22}$$

The dimensionality of the controllable subspace, spanned by the column space[2] of $\mathcal{C}$, (Im$\mathcal{C}$), is

$$\text{rank}\ \mathcal{C} = m \tag{2.23}$$

regardless of the number of landmarks in the map. Obviously, the only controllable states are the ones associated with the vehicle motion.

On the other hand, the observability matrix of our error driven system is

$$\mathcal{O} = \begin{bmatrix}\mathbf{H}\mathbf{F} \\ \mathbf{H}\mathbf{F}^2 \\ \vdots \\ \mathbf{H}\mathbf{F}^{m+dn}\end{bmatrix} \tag{2.24}$$

[2] See Appendix B for a formal definition of the four fundamental subspaces of linear algebra.

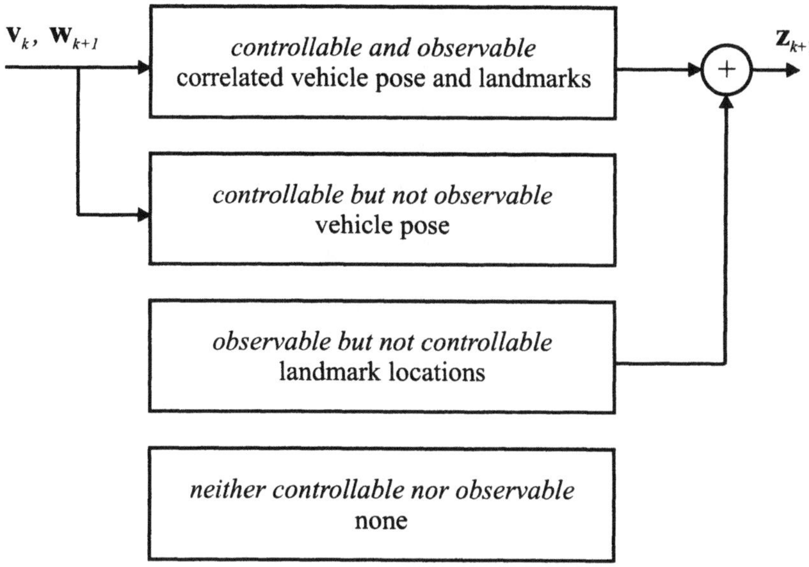

Fig. 2.3. Controllability and observability of the state space in CML.

The rank of $\mathcal{O}$ indicates the dimensionality of the observable subspace, which in turn, is spanned by the row space of $\mathcal{O}$, $(\text{Im}\mathcal{O}^\top)$.

$$\text{rank } \mathcal{O} = m + d(n-1) \tag{2.25}$$

The decomposition of the state space is portrayed in Figure 2.3. The arrows indicate their roles in the realization of the system.

2.5 The Monobot

We return our attention now to the simple linear 1-D robot from Section 1.2. Consider the even more restrictive case in which only one landmark is available. By substituting the resulting expressions for the model Jacobians in Equations 1.52 and 1.53, the controllability and observability matrices reduce to

$$\mathcal{C} = \begin{bmatrix} 1\,0\,0\,1\,0\,0 \\ 0\,0\,0\,0\,0\,0 \end{bmatrix} \tag{2.26}$$

$$\mathcal{O} = \begin{bmatrix} -1\;1 \\ -1\;1 \end{bmatrix} \tag{2.27}$$

The controllable subspace has a basis of the form $(q, 0)^\top$, clearly indicating that the only dimension in the state space that can be controlled is the one associated with the motion of the robot.

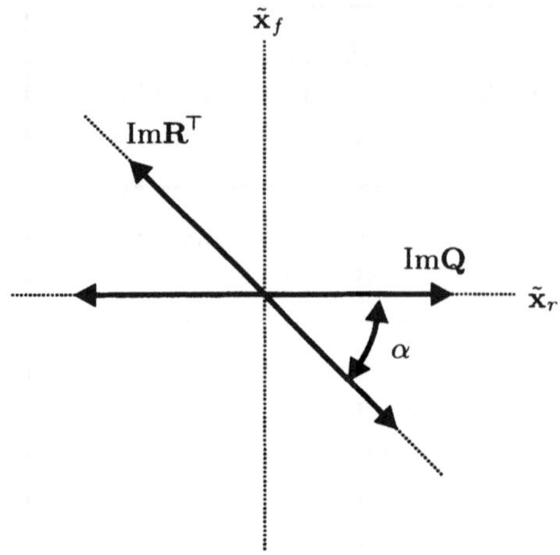

Fig. 2.4. Controllability and observability of a linear one dimensional mobile robot with one landmark.

The observable subspace on the other hand, with basis $(r, -r)^\top$, shows how the observed robot and landmark locations are fully correlated. This situation is portrayed graphically in Figure 2.4.

The unobservable subspace is the orthogonal complement of $\mathrm{Im}\mathcal{O}^\top$, and has a basis $[r, r]^\top$. An expression for it was already derived from the analysis of the total Fisher information matrix and is given in Equation 2.11. The name $\mathbf{E}_{\mathrm{Ker}\mathcal{O}}$ indicates that it is a basis for the null space of $\mathcal{O}$.

The controllable and observable subspaces for the one-landmark 1-D SLAM problem span along a pair of non-orthogonal lines in $\mathbb{R}^2$; with the consequence that the innovations only provide information for a fully correlated reconstruction of the state space.

We suggest to use the angle between these two subspaces, as a measure of the error incurred while trying to reconstruct the state $\tilde{\mathbf{x}}_r$ from correlated observations. It is evident from Figure 2.4 and Equations 2.26-2.27, that for the one landmark *monobot*, the angle is

$$\begin{aligned} \alpha &= \angle\, \mathrm{Im}\mathcal{C}\ \mathrm{Im}\mathcal{O}^\top \\ &= \pi/4\mathrm{rad}. \end{aligned} \tag{2.28}$$

To see the physical interpretation of the angle α in Equation 2.28, we ask the reader to analyze Figure 2.4 in detail once more. The error between an observed landmark and its prediction must lay along $\mathcal{O}^\top$. Such vector quantity, multiplied by its corresponding Kalman gain, is used to revise our estimates of both the robot and landmark locations. However, there is one direction of the state space which is not observed, the one orthogonal to $\mathrm{Im}\mathcal{O}^\top$ (along $\mathrm{Ker}\mathcal{O}$). The information for the revision of $\tilde{\mathbf{x}}_r$ and $\tilde{\mathbf{x}}_f$ along the direction orthogonal to $\mathrm{Im}\mathcal{O}^\top$ is missing. The angle α indicates how close noise driven observations are from fully revising the robot part of the state space.

The immediate questions that come to our mind are: what happens if we add more landmarks to the environment? Will the vehicle and landmark location estimates improve or degrade? Will we be able to achieve an uncoupled reconstruction of the entire state space?

Surprisingly enough (and we believe, the entire research community which supports the EKF-based approaches to SLAM will find these observations crucial), the answer to the above questions is "improve" but "no".

We will analyze now the two-landmark *monobot* case, and after that, we will introduce and prove a new closed form expression that links the degree of reconstruction in the EKF-SLAM algorithm to the number of landmarks used in the case of the *monobot.*

The observation Jacobian matrix for the two-landmark *monobot* case becomes

$$\mathbf{H} = \begin{bmatrix} -1\,1\,0 \\ -1\,0\,1 \end{bmatrix} \tag{2.29}$$

and the observability matrix of our noise driven linear state system is now

$$\mathcal{O} = \begin{bmatrix} -1\,1\,0 \\ -1\,0\,1 \\ -1\,1\,0 \\ -1\,0\,1 \\ -1\,1\,0 \\ -1\,0\,1 \end{bmatrix} \tag{2.30}$$

A possible set of bases for the controllable and observable subspaces are

$$\mathbf{E}_{\mathrm{Im}\mathcal{C}} = \begin{pmatrix} 1 \\ 0 \\ 0 \end{pmatrix} \tag{2.31}$$

$$\mathbf{E}_{\mathrm{Im}\mathcal{O}^\top} = \begin{pmatrix} 1 & 1 \\ -1 & 0 \\ 0 & -1 \end{pmatrix} \tag{2.32}$$

and the angle between these two subspaces can be computed as the smallest non null singular value of the product of their orthonormal bases [74, 87].

$$\mathbf{E}_{\mathrm{Im}\mathcal{C}} = \mathbf{U}_\mathcal{C}\boldsymbol{\Sigma}_\mathcal{C}\mathbf{V}_\mathcal{C}^\top \tag{2.33}$$

$$\mathbf{U}_\mathcal{C} = \begin{bmatrix} 1 \\ 0 \\ 0 \end{bmatrix} \tag{2.34}$$

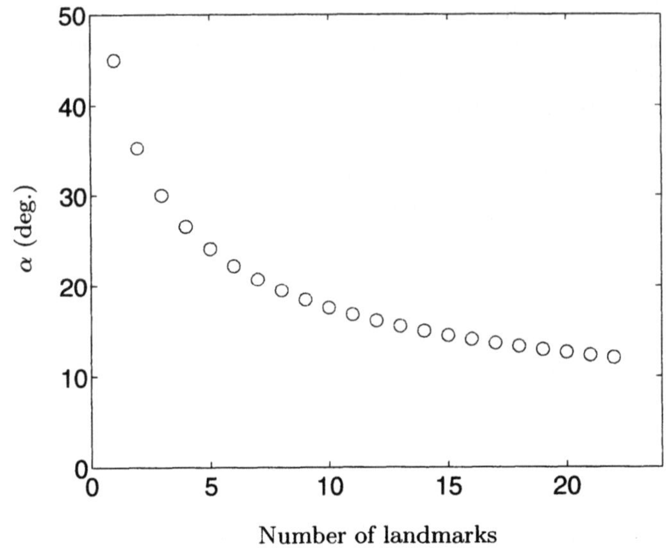

Fig. 2.5. $\angle \text{Im}(\mathcal{C})\text{Im}(\mathcal{O}^\top)$. Angle between the observable and controllable subspaces.

$$\mathbf{E}_{\text{Im}\mathcal{O}^\top} = \mathbf{U}_\mathcal{O}\boldsymbol{\Sigma}_\mathcal{O}\mathbf{V}_\mathcal{O}^\top \tag{2.35}$$

$$\mathbf{U}_\mathcal{O} = \begin{bmatrix} -0.7071 & 0.4082 \\ 0 & -0.8165 \\ 0.7071 & 0.4082 \end{bmatrix} \tag{2.36}$$

$$\begin{aligned} \alpha &= \angle \ \text{Im}\mathcal{C} \ \text{Im}\mathcal{O}^\top \\ &= \arccos \sigma_{\min}(\mathbf{U}_\mathcal{C}^\top \mathbf{U}_\mathcal{O}) \\ &= 163\pi/832\text{rad} \\ &\approx 0.6155\text{rad}. \end{aligned} \tag{2.37}$$

The decrease in the angle between the controllable and observable subspaces obtained by adding one more landmark to the map, suggests that our measurement noise driven corrections to the map state estimate would reconstruct the vehicle localization estimate closer to the actual value of the vehicle pose.

Following this procedure we computed the value of α for a three-landmark *monobot* model, further reducing to $\alpha = \pi/6$. And, as we add more landmarks to the map, the angle between the observable and controllable subspaces reduces monotonically. Figure 2.5 shows experimentally the decrease in α as landmarks are added to the map state model. We are ready to introduce the following theorem:

Theorem 2.1. *The angle between the controllable and observable subspaces in the case of a linear one-dimensional robot in the EKF-SLAM algorithm depends only on the total number of landmarks used (n), and is given by*

$$\alpha = \arccos \sqrt{\frac{n}{n+1}}$$

Proof. Carefully following the pattern just described for the computation of α; in a generalization for the n-landmark *monobot* case, a set of possible bases for the controllable and observable subspaces are given by

$$\begin{aligned} \mathbf{E}_{\mathrm{Im}\mathcal{C}} &= (\mathbf{q}) \\ &= \begin{pmatrix} 1 \\ \mathbf{0}_{n\times 1} \end{pmatrix} \end{aligned} \tag{2.38}$$

and

$$\begin{aligned} \mathbf{E}_{\mathrm{Im}\mathcal{O}^\top} &= \begin{pmatrix} \mathbf{r}_1 \ldots \mathbf{r}_n \end{pmatrix} \\ &= \begin{pmatrix} \mathbf{1}_{1\times n} \\ -\mathbf{I} \end{pmatrix} \end{aligned} \tag{2.39}$$

respectively.

Moreover, let $\mathbf{p}$ be the projection of $\mathbf{E}_{\mathrm{Im}\mathcal{C}}$ onto $\mathbf{E}_{\mathrm{Im}\mathcal{O}^\top}$. $\mathbf{p}$ is easily computed as the sum of the individual projections of $\mathbf{q}$ onto each element $\mathbf{r}_i$ of the basis of the observable space, i.e.,

$$\mathbf{p} = \sum_{i=1}^{n} \frac{\mathbf{q}^\top \mathbf{r}_i}{\mathbf{r}_i^\top \mathbf{r}_i} \mathbf{r}_i \tag{2.40}$$

Substituting the basis of the controllable space in Equation 2.40 gives the reduced expression

$$\begin{aligned} \mathbf{p} &= \frac{1}{2} \sum_{i=1}^{n} \mathbf{r}_i \\ &= \frac{1}{2} \begin{bmatrix} n \\ -\mathbf{1}_{n\times 1} \end{bmatrix} \end{aligned} \tag{2.41}$$

Finally, the angle between $\mathbf{p}$ and $\mathbf{q}$, and consequently between the two subspaces, is

$$\begin{aligned} \alpha &= \arccos \frac{\mathbf{p}^\top \mathbf{q}}{\|\mathbf{p}\| \|\mathbf{q}\|} \\ &= \arccos \sqrt{\frac{n}{n+1}} \end{aligned} \tag{2.42}$$

As the number of landmarks grows, the observable subspace gets closer to the controllable part of the state space (the vehicle localization states).

$$\begin{aligned} \lim_{n\to\infty} \alpha &= \lim_{n\to\infty} \arccos \sqrt{\frac{n}{n+1}} \\ &= 0 \end{aligned} \qquad (2.43)$$

It is unrealistic however, to have an infinite number of landmarks, and a compromise has to be made between the possibility of including as many landmarks as possible, and the amount of information that new observations give. Also one has to bear in mind that as we add more and more landmarks to the map, we will also introduce their associated measurement noise.

Consequently, it has been argued that the performance of the SLAM algorithm would be enhanced by concentrating on fewer, better landmark observations [41]. And that is certainly true, little gain (little reduction of α) is made from going from 25 to 125 landmarks compared to the move from 1 to 5 or 5 to 25.

In Figures 2.6-2.7 we have plotted the results of using the original fully correlated approach to SLAM for a monobot that starts at location $\mathbf{x}_{r,0|0} = -1m$, and moves along a straight line with a temporal sinusoid trajectory returning to the same point after 100 iterations. Landmarks are located at $\mathbf{x}_{f(i)} = 1m$. A plant noise model with σ_v proportional to the motion command by 10%, and a measurement noise model with σ_w proportional to the distance from the sensor to the landmark by 1% were used.

In the first column in Figure 2.6, the solid lines indicate the true and estimated state trajectories, as well as the actual landmark measurements. The blue dotted line shows 2σ bounds on the state estimate.

The effect of partial observability manifests itself in the dependence on the initial conditions. Note how both the vehicle and landmark mean localization errors do not converge to zero. Their steady state value is subject to the error incurred at the first observation. That is, the filter is unstable.

A Montecarlo simulation over 100 SLAM runs showed however filter unbiasedness, a property of optimal stochastic state estimation (Kalman filter). That is, the average landmark localization error over the entire set of simulations was still zero, thanks to the independence of the initial landmark measurement errors at each test run.

Partial controllability on the other hand, produces a zero Kalman gain for the revision of the landmark estimates. That is, after a few iterations the Kalman filter believes it has a perfectly accurate estimate of the landmark locations, contradictory to the localization error just described. The rate at which the landmark localization Kalman gain approaches zero is dictated by the rate of convergence of the system, i.e., the system's time constant (see the bibliographical notes).

Moreover, the steady state error for the robot and landmark localization is less sensitive to the initial conditions when a large number of landmarks

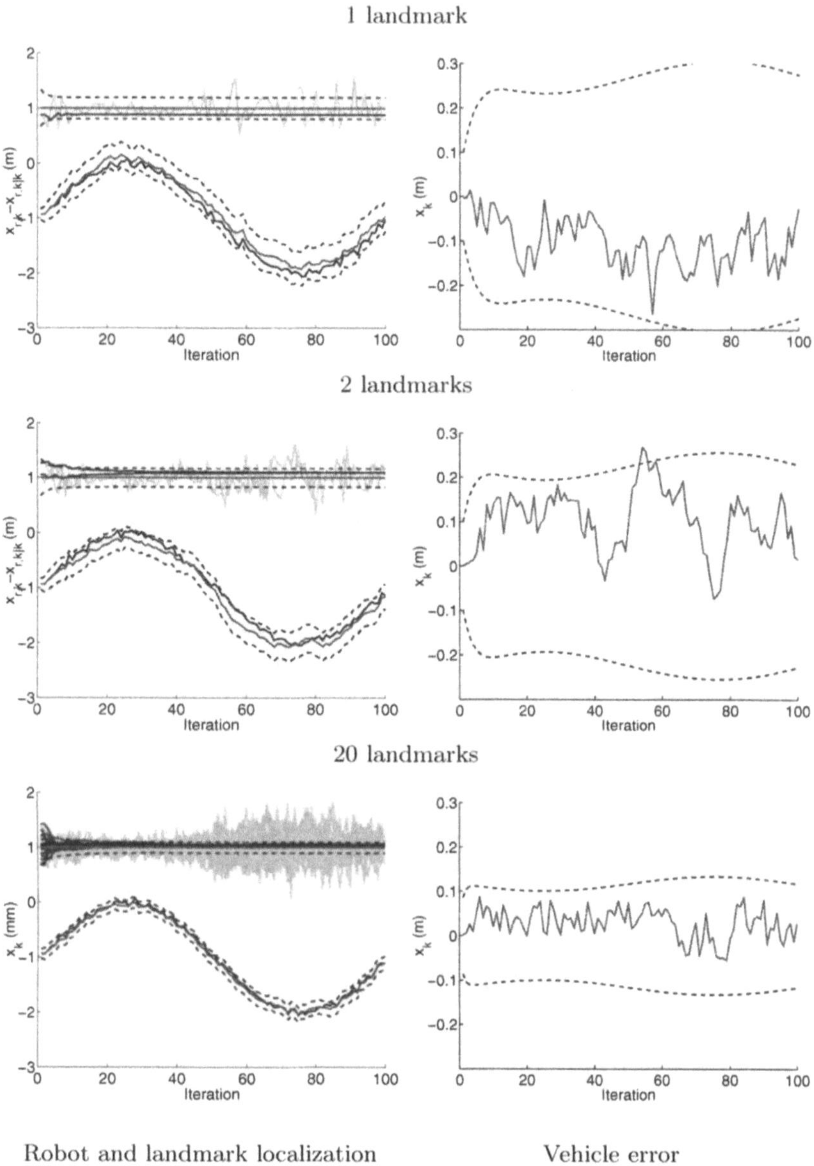

Fig. 2.6. Full-covariance EKF SLAM for monobot in a sinusoidal path starting at $\mathbf{x}_{r,0|0} = -1m$, and 100 iterations

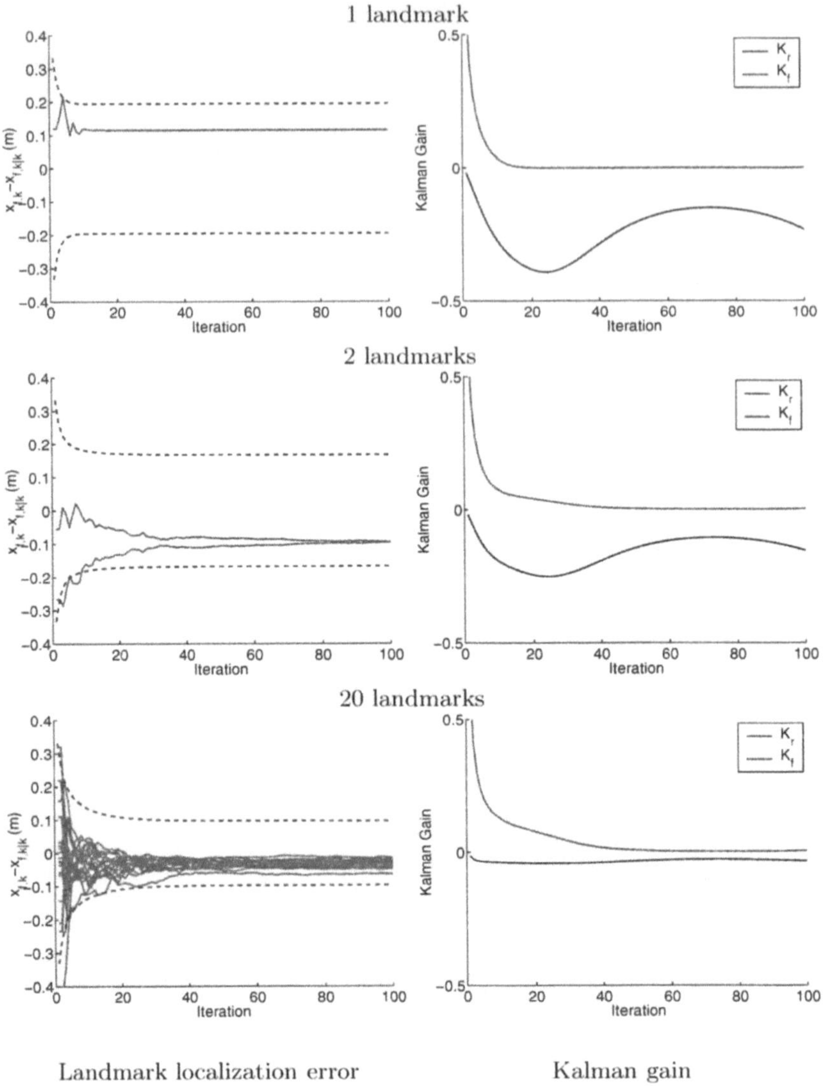

Fig. 2.7. Full-covariance EKF SLAM for monobot in a sinusoidal path starting at $\mathbf{x}_{r,0|0} = -1m$, and 100 iterations.

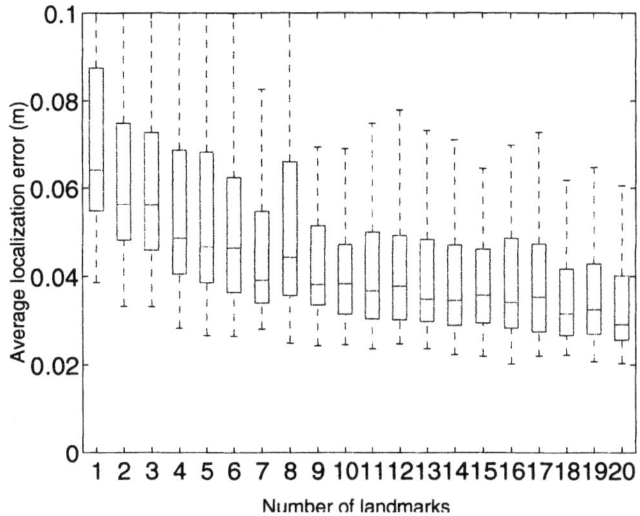

Fig. 2.8. Reduction of the average monobot localization error $\mathbf{x}_{r,k} - \mathbf{x}_{r,k|k}$ with respect to the number of landmarks used. The results correspond to a Montecarlo simulation over 100 SLAM runs. The dotted lines show the extent of the data for the entire set of runs, and the boxes contain marks at the lower, median and upper quartile.

are used. The reason is the same as for the Montecarlo simulation, the observations are independent, and their contribution averages at each iteration in the computation of the localization estimate. The results of the Montecarlo simulation are shown in Figure 2.8 depicting the effect of the increase in the number of landmarks on the average vehicle localization error.

Theorem 2.1 about the amount of reconstruction possible in EKF-SLAM with respect to the number of landmarks applies only to the simple *monobot* case. We will concentrate our attention now on a more realistic case, a planar mobile vehicle.

2.6 The Planar Robot

We will show now how the reconstructibility issues presented for the linear and one-dimensional robot of the previous section, nicely extend when studying more complicated platforms. We investigate now the case of the planar robot presented in Section 1.2, a non-holonomic nonlinear wheeled vehicle with 3 DOF, and an environment consisting of 2-D point landmarks located on the floor.

The dimensionality of the controllable subspace is $m = 3$, and for the specific case in which only one landmark is available, a basis for the controllable subspace is simply

$$\mathbf{E}_{\mathrm{Im}\mathcal{C}} = \begin{pmatrix} \mathbf{I} \\ \mathbf{0}_{2\times 3} \end{pmatrix}$$

The dimensionality of the observable space is, for this particular configuration, rank $\mathcal{O} = 3$. This last result is easily verified with simple symbolic manipulation of the specific expression for the state model from Section 1.2. Furthermore, one possible basis for $\mathrm{Im}\mathcal{O}^\top$ is

$$\mathbf{E}_{\mathrm{Im}\mathcal{O}^\top} = \begin{pmatrix} 1 & 0 & 0 \\ 0 & 1 & 0 \\ 0 & 0 & 1 \\ -1 & 0 & 0 \\ 0 & -1 & 0 \end{pmatrix}$$

Our 3 DOF robot along with a 2-D landmark form a map state space in $\mathbb{R}^5$, and the null space of $\mathcal{O}$ (the unobservable subspace) is spanned by

$$\mathbf{E}_{\mathrm{Ker}\mathcal{O}} = \begin{pmatrix} 1 & 0 \\ 0 & 1 \\ 0 & 0 \\ 1 & 0 \\ 0 & 1 \end{pmatrix}$$

Recall from the basic properties of the four fundamental subspaces of linear algebra that $\mathrm{Im}\mathcal{O}^\top + \mathrm{Ker}\mathcal{O} = \mathbb{R}^5$, $\mathrm{Im}\mathcal{O}^\top \cap \mathrm{Ker}\mathcal{O} = \emptyset$. The only independently observable state is the one associated to the robot orientation θ. The other four states, the Cartesian coordinates of the robot and landmark locations depend linearly on each other.

Moreover, even when $\mathrm{Im}\mathcal{C}$ and $\mathrm{Im}\mathcal{O}^\top$ both span $\mathbb{R}^3$, we see that the inequality $\mathrm{Im}\mathcal{C} \neq \mathrm{Im}\mathcal{O}^\top$ still holds, as in the case of the *monobot*. That is, the observable and controllable subspaces for the one-landmark 3 DOF robot SLAM problem correspond to different 3-D subspaces in $\mathbb{R}^5$; and, their intersection represents the only fully controllable and observable state, i.e., the robot orientation. A pictorial representation of this situation is depicted in Figure 2.9, and once more, a measure of the reconstruction error incurred when estimating the vehicle pose from correlated observations is given by the angle between these two subspaces.

Resorting again to the singular value decomposition for the computation of a pair of orthonormal bases for $\mathrm{Im}\mathcal{C}$ and $\mathrm{Im}\mathcal{O}^\top$, we have that for the one-landmark planar robot case

$$\mathbf{E}_{\mathrm{Im}\mathcal{C}} = \mathbf{U}_\mathcal{C} \boldsymbol{\Sigma}_\mathcal{C} \mathbf{V}_\mathcal{C}^\top \tag{2.44}$$

$$\mathbf{U}_\mathcal{C} = \begin{pmatrix} \mathbf{I} \\ \mathbf{0}_{2\times 3} \end{pmatrix} \tag{2.45}$$

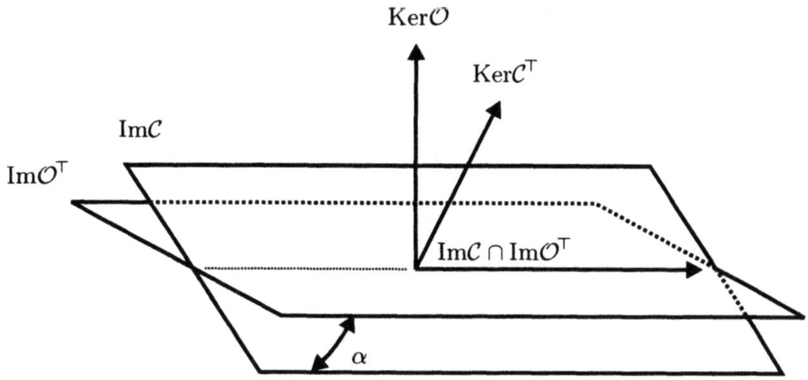

Fig. 2.9. Controllability and observability subspaces.

$$\mathbf{E}_{\mathrm{Im}\mathcal{O}^\top} = \mathbf{U}_\mathcal{O}\boldsymbol{\Sigma}_\mathcal{O}\mathbf{V}_\mathcal{O}^\top \tag{2.46}$$

$$\mathbf{U}_\mathcal{O} = \begin{pmatrix} -0.7071 & 0 & 0 \\ 0 & -0.7071 & 0 \\ 0 & 0 & -1 \\ 0.7071 & 0 & 0 \\ 0 & 0.7071 & 0 \end{pmatrix} \tag{2.47}$$

and, as before,

$$\alpha = \angle\, \mathrm{Im}\mathcal{C}\ \mathrm{Im}\mathcal{O}^\top \tag{2.48}$$

$$= \arccos \sigma_{\min}(\mathbf{U}_\mathcal{C}^\top \mathbf{U}_\mathcal{O}) \tag{2.49}$$

$$= \pi/4\mathrm{rad} \tag{2.50}$$

Note the equivalence of α in Equations 2.28 and 2.50. The immediate inference behind this result is that neither the nonlinearities of the planar mobile robot platform, nor the change in the dimensionality of the model, are related to the amount of reconstruction possible when using a fully correlated SLAM model.

We now elaborate more on this, because if we can show this result to hold for a map with more landmarks, we would have a powerful argument to defend the hypothesis that the number of landmarks used in SLAM is directly related to the average localization error in the form of Figures 2.5 and 2.8.

Extending now our study to the two-landmark case; possible orthonormal bases for the controllable and observable subspaces as computed with the aforementioned *svd* are given by

$$\mathbf{U}_\mathcal{C} = \begin{pmatrix} \mathbf{I} \\ \mathbf{0}_{4\times 3} \end{pmatrix}$$

$$\mathbf{U}_{\mathcal{O}} = \begin{pmatrix} -0.4082 & 0 & 0 & 0 & 0.7071 \\ 0 & 0.4082 & 0 & 0.7071 & 0 \\ 0 & 0 & -1 & 0 & 0 \\ -0.4082 & 0 & 0 & 0 & -0.7071 \\ 0 & 0.4082 & 0 & -0.7071 & 0 \\ 0.8165 & 0 & 0 & 0 & 0 \\ 0 & -0.8165 & 0 & 0 & 0 \end{pmatrix}$$

and once more, the angle between them is

$$\alpha = \arccos \sigma_{\min}(\mathbf{U}_{\mathcal{C}}^{\top}\mathbf{U}_{\mathcal{O}}) \tag{2.51}$$

$$\alpha = 163\pi/832. \tag{2.52}$$

Similarly, for a three-landmark model, $\alpha = \pi/6$, and as we add more and more landmarks to the environment, the angle between the controllable and observable subspaces reduces monotonically, in exactly the same manner as in the case of the *monobot.* We are now ready to formulate the second theorem in this Chapter

Theorem 2.2. *The angle between the controllable and observable subspaces in the case of a nonlinear planar robot with 3 degrees of freedom in the EKF-SLAM algorithm, depends only on the total number of landmarks used (n), and is given by*

$$\alpha = \arccos \sqrt{\frac{n}{n+1}}$$

Proof. The key to the proof is in Figure 2.9. Notice that thanks to the orthogonality of the four fundamental subspaces, the angle between the observable and controllable subspaces is exactly the same as the angle between their complementary subspaces. This is,

$$\alpha = \angle\, \mathrm{Ker}\mathcal{C}^{\top}\ \mathrm{Ker}\mathcal{O} \tag{2.53}$$

The controllable subspace has a fixed rank of size $m = 3$, regardless of the number of landmarks; and the size of the basis for the observable subspace would depend on n. Now, the roles are reversed. The dimension of $\mathbf{E}_{\mathrm{Ker}\mathcal{C}^{\top}}$ grows with respect to the number of landmarks, but maintains a very simple form

$$\mathbf{E}_{\mathrm{Ker}\mathcal{C}^{\top}} = \begin{bmatrix} \mathbf{0}_{3\times 2n} \\ \mathbf{I} \end{bmatrix} \tag{2.54}$$

The null complement of the observable subspace on the other hand has a fixed number of columns (just two), and it can be easily shown by inspection that

$$\mathbf{E}_{\mathrm{Ker}\mathcal{O}} = \begin{bmatrix} \mathbf{I} \\ \mathbf{0}_{1\times 2} \\ \mathbf{I} \\ \vdots \\ \mathbf{I} \end{bmatrix}_{(3+2n)\times 2} \tag{2.55}$$

These are precisely the directions along which our state space is unobservable. Clearly showing that in the EKF-SLAM model, the Cartesian coordinates of the robot and landmark locations are all fully correlated.

The angle between these two subspaces is again, given by the smallest singular value of an orthonormalized version of the product $\mathbf{E}_{\text{Ker}\mathcal{C}^\top}{}^\top \mathbf{E}_{\text{Ker}\mathcal{O}}$, in which $\mathbf{E}_{\text{Ker}\mathcal{C}^\top} = \mathbf{U}_\mathcal{C} \boldsymbol{\Sigma}_\mathcal{C} \mathbf{V}_\mathcal{C}^\top$, and $\mathbf{E}_{\text{Ker}\mathcal{O}} = \mathbf{U}_\mathcal{O} \boldsymbol{\Sigma}_\mathcal{O} \mathbf{V}_\mathcal{O}^\top$ are the singular value decompositions of $\mathbf{E}_{\text{Ker}\mathcal{C}^\top}$ and $\mathbf{E}_{\text{Ker}\mathcal{O}}$, respectively. This is,

$$\mathbf{U}_\mathcal{C}^\top \mathbf{U}_\mathcal{O} = \frac{1}{\sqrt{n+1}} \begin{bmatrix} \mathbf{I} \\ \vdots \\ \mathbf{I} \end{bmatrix}_{2\times 2n} \tag{2.56}$$

$$\alpha = \arccos \sigma_{\min}(\mathbf{U}_\mathcal{C}^\top \mathbf{U}_\mathcal{O}) \tag{2.57}$$

$$\alpha = \arccos \sqrt{\frac{n}{n+1}} \tag{2.58}$$

Asymptotic stability of the KF means that its solution will gradually become insensitive to its initial conditions. One can see that observability plays a role because, if there are sufficient landmark measurements, the true localization and landmark estimates will be well approximated. Also, controllability will play a role because if the system is not controllable in some modes (as it is), then any number of observations cannot help damp the analysis errors.

2.7 Observability

In Section 2.2 we characterized the unobservable subspace in SLAM as the subspace spanned by the null eigenvectors of the total Fisher information matrix. Furthermore, we showed in Sections 2.4-2.6 how the unobservable part of the state space is precisely a linear combination of the landmark and robot pose estimates.

In order to gain full observability we propose to extend the measurement model doing away with the constraint imposed by full correlation. We present two techniques to achieve this. One is to let one landmark serve as a fixed global reference, with its localization uncertainty independent to the vehicle pose.

The second proposed technique is the addition of a fixed external sensor, such as a camera or a GPS, that can measure the vehicle location at all times, independent of the landmark estimates.

Both techniques are based essentially on the same principle. Full observability requires an uncorrelated measurement Jacobian, or equivalently, a full rank Fisher information matrix.

We next present, without loss of generality, the extensions to the monobot SLAM model in order to obtain full observability.

A fixed global reference

The plant model is left untouched, i.e., (from Equations 1.1 and 1.50)

$$\mathbf{x}_{k+1} = \mathbf{x}_k + \mathbf{u}_k + \mathbf{v}_k \tag{2.59}$$

The measurement model takes now the form

$$\begin{bmatrix} z_k^{(0)} \\ \mathbf{z}_k \end{bmatrix} = \begin{bmatrix} -1 & \mathbf{0}_{1\times n} \\ -\mathbf{1}_{n\times 1} & \mathbf{I} \end{bmatrix} \mathbf{x} + \begin{bmatrix} w_k^{(0)} \\ \mathbf{w}_k \end{bmatrix} \tag{2.60}$$

One of the observed landmarks is to be taken as a global reference at the world origin. No map state is needed for it. The zero-th superscript in the measurement vector is used for the consistent indexing of landmarks and observations with respect to the original model. It can be easily shown that the observability matrix for this new model is full rank.

The innovation covariance matrix for the augmented system is of size $n+1 \times n+1$, and unlike Equation 2.8, its corresponding Fisher information matrix is of full rank.

Figures 2.10 and 2.11 show the results of applying a full observability to the same monobot model as the one portrayed in Figures 2.6-2.7. Note how the steady state (robot pose and landmark locations) is now unbiased with respect to the initial landmark estimates. Furthermore, state covariances are also smaller than those in Figures 2.6-2.7.

An external sensor

Instead of using one of the landmarks as a global reference, one could also use a fixed sensor to measure the position of the robot. For example, by positioning a camera that observes the vehicle at all times. For such cases, the monobot measurement model may take the form

$$\begin{bmatrix} z_k^{(0)} \\ \mathbf{z}_k \end{bmatrix} = \begin{bmatrix} 1 & \mathbf{0}_{1\times n} \\ -\mathbf{1}_{n\times 1} & \mathbf{I} \end{bmatrix} \mathbf{x} + \begin{bmatrix} w_k^{(0)} \\ \mathbf{w}_k \end{bmatrix} \tag{2.61}$$

The characteristics on the observability matrix, and the Fisher information matrix, are exactly the same as for the previous case. This new model is once more, fully observable. Figures 2.12 and 2.13 show the results of using an external sensor to measure the vehicle pose. The results are theoretically equivalent to the previous case. The choice of one technique over the other one would depend on the availability of such external sensor, and on its measurement noise covariance characteristics.

The key point here is that we have proved that full observability, i.e., zero mean steady state error, is indeed possible in SLAM without the need of an oracle (an external sensor) whose reading needs to be available at all times in order to preserve observability, but by simply anchoring the first observed landmark to the global reference frame.

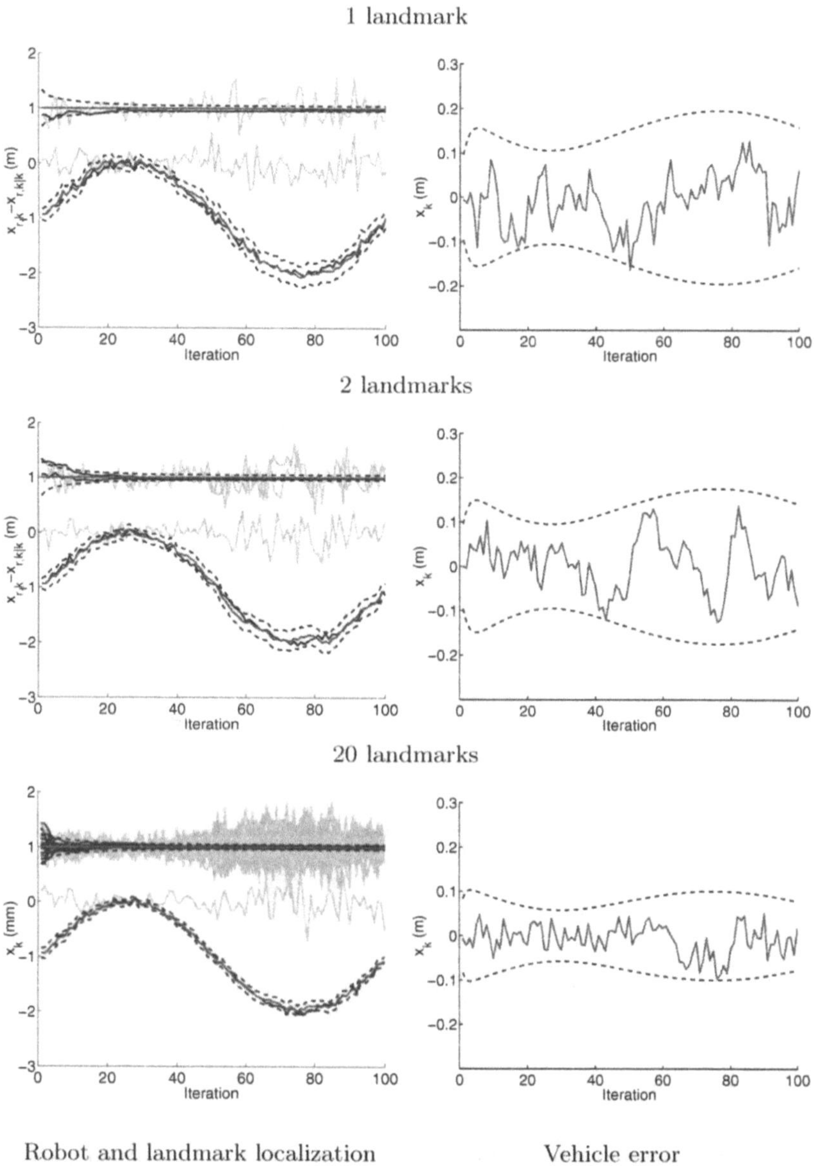

Fig. 2.10. Full-covariance fully observable EKF SLAM for monobot in a sinusoidal path starting at $\mathbf{x}_{r,0|0} = -1m$, and 100 iterations. The global reference is observed at the origin.

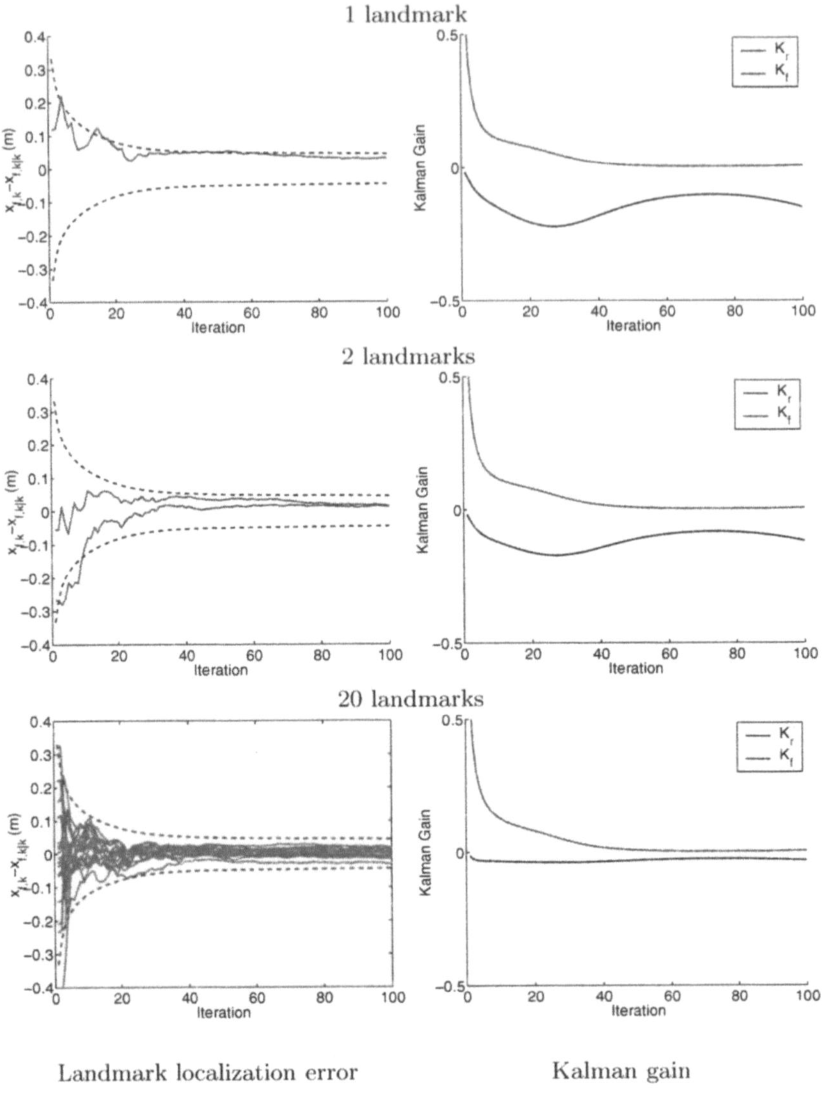

Fig. 2.11. Full-covariance fully observable EKF SLAM for monobot in a sinusoidal path starting at $\mathbf{x}_{r,0|0} = -1m$, and 100 iterations. The global reference is observed at the origin.

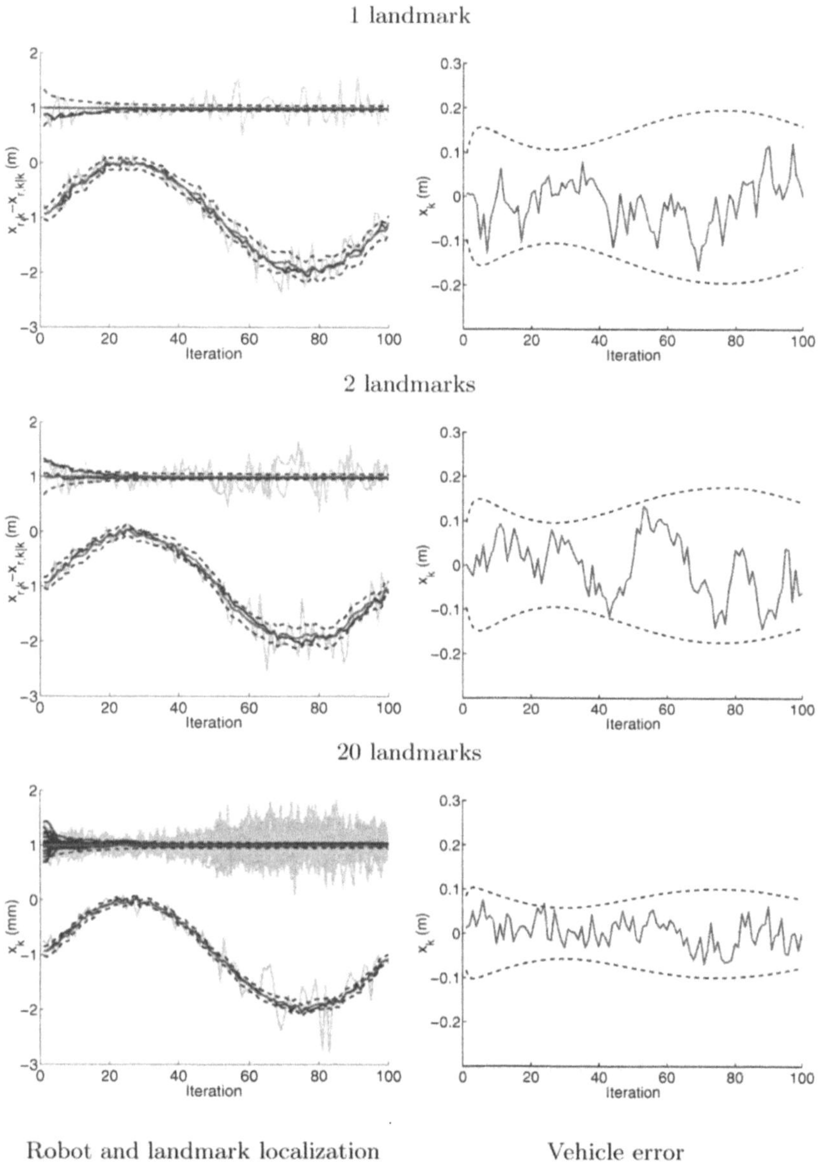

Fig. 2.12. Full-covariance fully observable EKF SLAM for monobot in a sinusoidal path starting at $\mathbf{x}_{r,0|0} = -1m$, and 100 iterations. The global reference is observed at the origin. A fixed external sensor is used for the measurement of the vehicle pose.

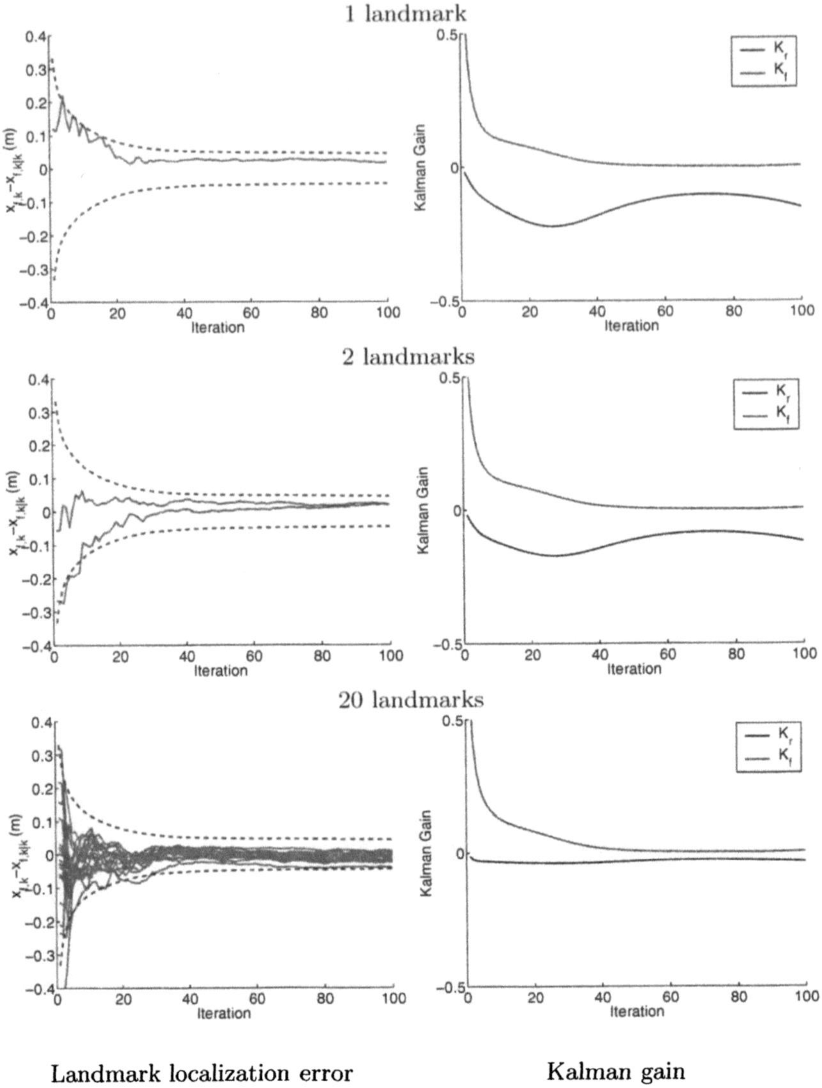

Fig. 2.13. Full-covariance fully observable EKF SLAM for monobot in a sinusoidal path starting at $\mathbf{x}_{r,0|0} = -1m$, and 100 iterations. The global reference is observed at the origin. A fixed external sensor is used for the measurement of the vehicle pose.

As with any sequential innovation KF for a fully observable system, partial observability occurs until the entire set of observations spanning the state space is completed. So, if the anchor landmark is not observed for a certain period of time, the filter will be reconstructing a partially observable state estimate. But, when the anchor landmark is re-observed, the system becomes fully observable again.

Full observability however, cannot be guaranteed if the vehicle looses permanent sight of the initial anchor. Nevertheless, the effect of partial observability is precisely the steady state error (and larger covariance) due to coupled error at the first iteration of the filter. So, in practice, full observability need only be guaranteed at the beginning of the filter.

Finally, a good strategy for any local submap approach to SLAM is to build many local maps, one attached to each anchor needed to cover the entire mapped area. In this way, full observability will guarantee optimal vehicle and landmark localization, with the smallest possible variances for each submap.

Planar vehicle

The results from the previous section are easily extensible to more complicated vehicle models. For example, the measurement model of a global reference fixed at the origin, for the nonlinear vehicle from Figure 1.7 is

$$\mathbf{h}^{(0)} = \begin{bmatrix} \sqrt{x_k^2 + y_k^2} + w_{r,k} \\ \tan^{-1}\left(\frac{y_k}{x_k}\right) - \theta_k + \frac{\pi}{2} + w_{\beta,k} \end{bmatrix} \tag{2.62}$$

and its corresponding Jacobian is

$$\mathbf{H}_0 = \begin{bmatrix} \frac{x_k}{\sqrt{x_k^2+y_k^2}} & \frac{y_k}{\sqrt{x_k^2+y_k^2}} & 0 & 0\,0\ldots \\ -\frac{y_k}{x_k^2+y_k^2} & \frac{x_k}{x_k^2+y_k^2} & -1 & 0\,0 \end{bmatrix} \tag{2.63}$$

The symbolic manipulation of

$$\mathbf{H} = \begin{bmatrix} \mathbf{H}_0 \\ \mathbf{H}_i \end{bmatrix} \tag{2.64}$$

with a commercial algebra package, produces a full rank observability matrix. That is, for the linearized nonholonomic velocity controlled planar mobile robot platform used, the simultaneous measurement of one anchor as global reference, and any other landmark, is sufficient to attain full observability in SLAM.

We show now results on a couple of experiments with a nonlinear vehicle with an also nonlinear measurement model, using in this case, the ACFR - University of Sydney database [69]. The test run used corresponds to a car-like vehicle at the University Car Park. The landmarks used are tree trunks, as measured with a laser range finder. The reconstructed maps are compared to GPS ground truth for accuracy. The first experiment corresponds to a typical partially observable SLAM run. Figure 2.14 plots results on this run, column a) shows the actual vehicle path and landmark location estimates recovered by the algorithm, compared to GPS ground truth for the beacons. Columns b) and d) show the covariances both for the vehicle and landmark state estimates. Note that even when the "relative" map is consistent [31], it is slightly rotated and shifted from the actual beacon locations. The amount of this shift depends on the initial vehicle uncertainty, i.e., the initial filter conditions, and can be seen in Figure 2.14, column c).

The second experiment shown corresponds to a fully observable SLAM run (using the first observed beacon as an anchor [11]). In this case, the vehicle and landmark covariance estimates do not depend on the initial filter conditions, and thus are significantly reduced. This is shown in columns b) and d) in Figure 2.15. The absolute landmark estimate error is also significantly reduced, as shown in column c). Figure 2.16 shows the actual vehicle path and landmark estimates as recovered by the filter. The beacon shown in the center of the plot is used as an anchor to the map, and no state estimate is computed for it.

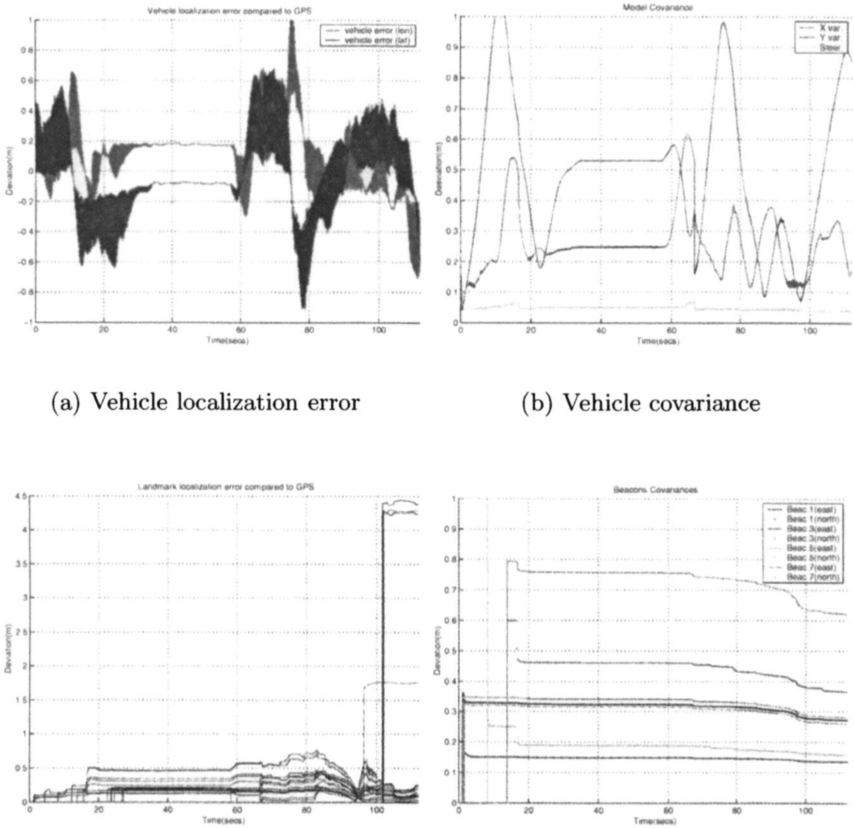

(a) Vehicle localization error (b) Vehicle covariance

(c) Landmark localization errors (d) Landmark covariance

Fig. 2.14. Partially observable SLAM for a car-like vehicle at the University of Sydney Car Park.

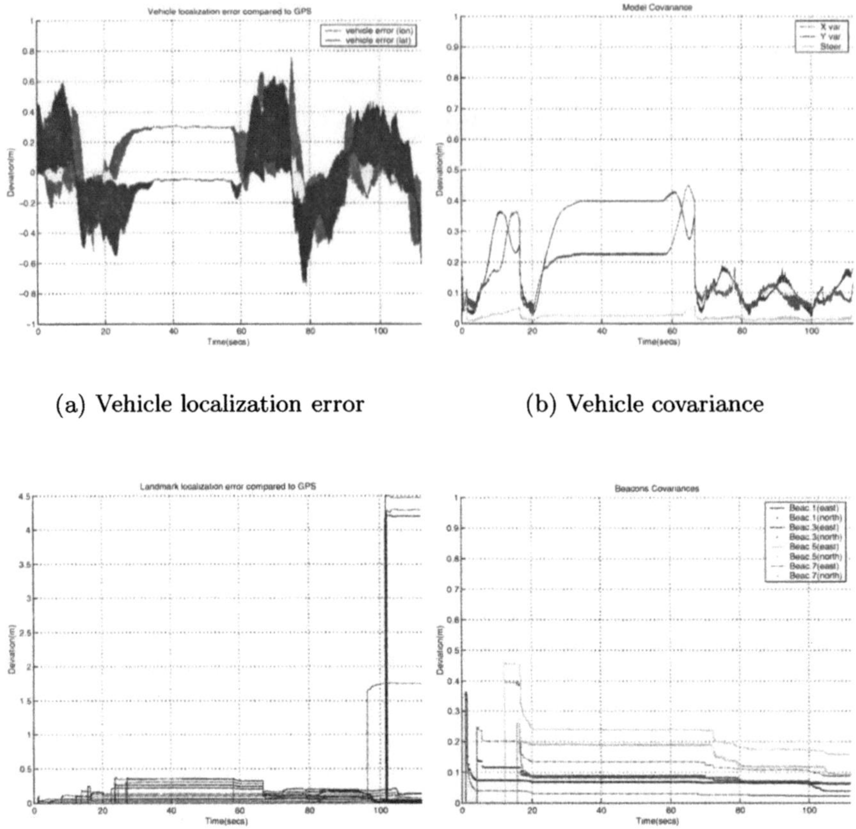

(a) Vehicle localization error

(b) Vehicle covariance

(c) Landmark localization errors

(d) Landmark covariance

Fig. 2.15. Fully observable SLAM for a car-like vehicle at the University of Sydney Car Park.

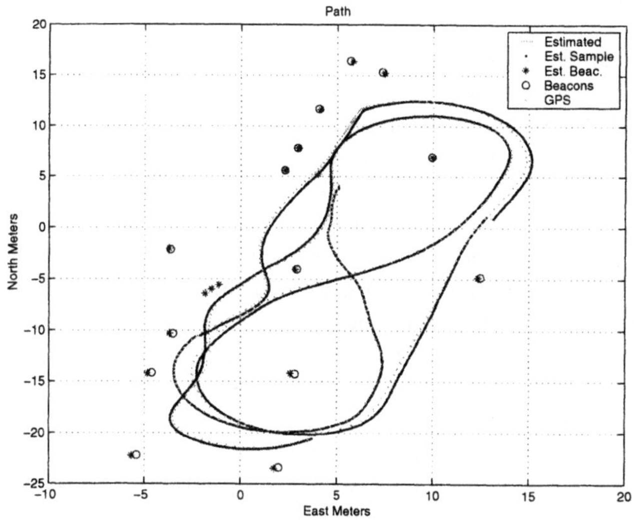

(a) Vehicle path and landmark localization

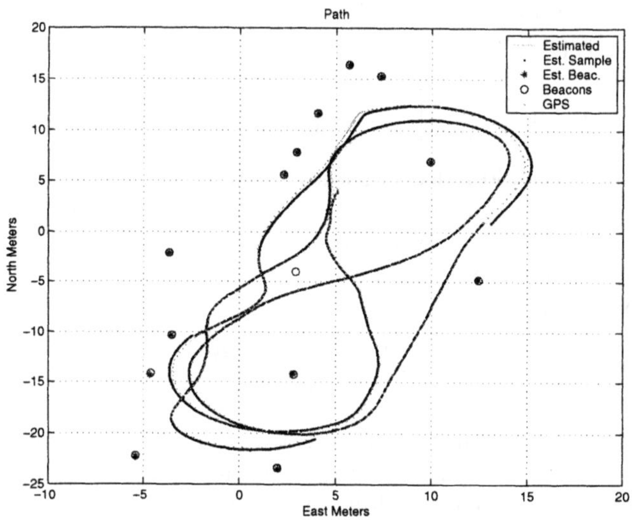

(b) Vehicle path and landmark localization

Fig. 2.16. Vehicle path and landmark location estimates, compared to GPS ground truth for an a) partially observable suboptimal SLAM run, and a b) fully observable suboptimal SLAM run.

2.8 Controllability

We have seen in Section 2.4 that in the standard SLAM case, the only controllable states are the ones associated to the vehicle location estimate.

Equation 1.12 assumes the landmarks are fixed elements, for which no process noise is considered. Therefore, their associated noise covariance (its determinant) will asymptotically tend to zero (see Section 1.1, Equation 1.25). The filter gain for the landmark states will also tend to zero.

Having a positive semidefinite covariance reflects the belief that there is a perfectly accurate estimate of some states (or linear combinations of them, the ones associated to the null eigenvalues of $\mathbf{P}$). This is, again, because there is no process noise entering these states, and the controllability condition described at the beginning of Section 2.4 pertaining the solution of the Riccati Equation 2.1 is not satisfied.

Since the fully correlated Kalman filter will not in general, yield consistent estimates for the map states (partial observability and divergence due to nonlinearities produce biased estimates); the situation where the map state covariance (its determinant) tends to zero is undesirable. It will lead in practice to smaller filter-computed variances than the estimated errors.

The situation can be remedied to some extent, by assuming the existence of artificial pseudo-noise associated to the landmark state estimates. This is, we can replace Equation 1.12 with

$$\begin{bmatrix} \mathbf{x}_{r,k+1} \\ \mathbf{x}_f \end{bmatrix} \approx \begin{bmatrix} \mathbf{x}_{r,k+1|k} \\ \mathbf{x}_{f,k|k} \end{bmatrix} + \begin{bmatrix} \mathbf{F}_r & \\ & \mathbf{I} \end{bmatrix} \begin{bmatrix} \tilde{\mathbf{x}}_{r,k|k} \\ \tilde{\mathbf{x}}_{f,k|k} \end{bmatrix} + \begin{bmatrix} \mathbf{G}_r & \\ & \mathbf{I} \end{bmatrix} \begin{bmatrix} \mathbf{v}_{r,k} \\ \mathbf{v}_{f,k} \end{bmatrix} \quad (2.65)$$

The landmark pseudo-noise $\mathbf{v}_{f,k}$ is assumed zero mean and white. Any positive definite covariance of this process noise will prevent the filter calculated covariance for the landmark states from converging to zero. The new process noise covariance is computed as the sum of the vehicle model covariance plus that of the pseudo noise.

$$\mathbf{Q}_{C,k} = \begin{bmatrix} \mathbf{Q}_{r,k} & \\ & \mathbf{0} \end{bmatrix} + \begin{bmatrix} \mathbf{0} & \\ & \mathbf{Q}_f \end{bmatrix} \quad (2.66)$$

The positive definite process noise covariance $\mathbf{Q}_{C,k}$ is substituted for $\mathbf{Q}_k$ in Equation 1.16. Figures 2.17 and 2.18 show the results of adding an artificial landmark state covariance to the fully observable model from Figures 2.10 and 2.13

Having a fully observable SLAM model with artificial process noise makes feasible the computation of the Cramer Rao lower bound. The monotonic dotted lines show the vehicle and landmark associated components of $\mathbf{J}^{-1}$. Moreover, Figure 2.19 shows the numerical evaluation of the eigenvalues for the matrix difference $\mathbf{P}_{k|k} - \mathbf{J}^{-1}$. When both eigenvalues are positive, the matrix difference is positive definite. That is, we can only trust our second order statistic $\mathbf{P}_{k|k}$ after a certain number of iterations have elapsed; only after enough information has been input to the filter in the form of innovations.

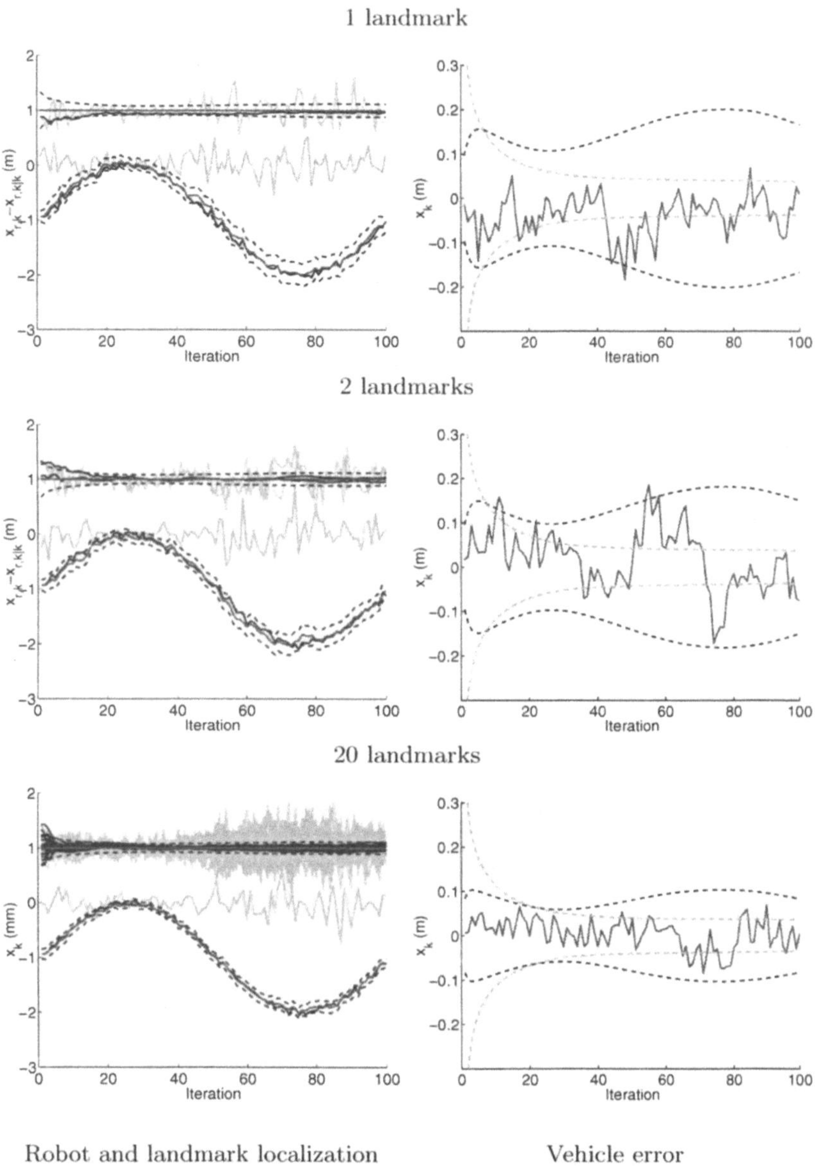

Fig. 2.17. Full-covariance fully observable SLAM with artificial process noise for a monobot in a sinusoidal path starting at $\mathbf{x}_{r,0|0} = -1m$, and with 100 iterations. A fixed external sensor is used for the measurement of the vehicle pose, and a pseudo-noise covariance ($\mathbf{V}_f = 0.05^2\mathbf{I}$) is used to prevent the landmark Kalman gains from converging to zero.

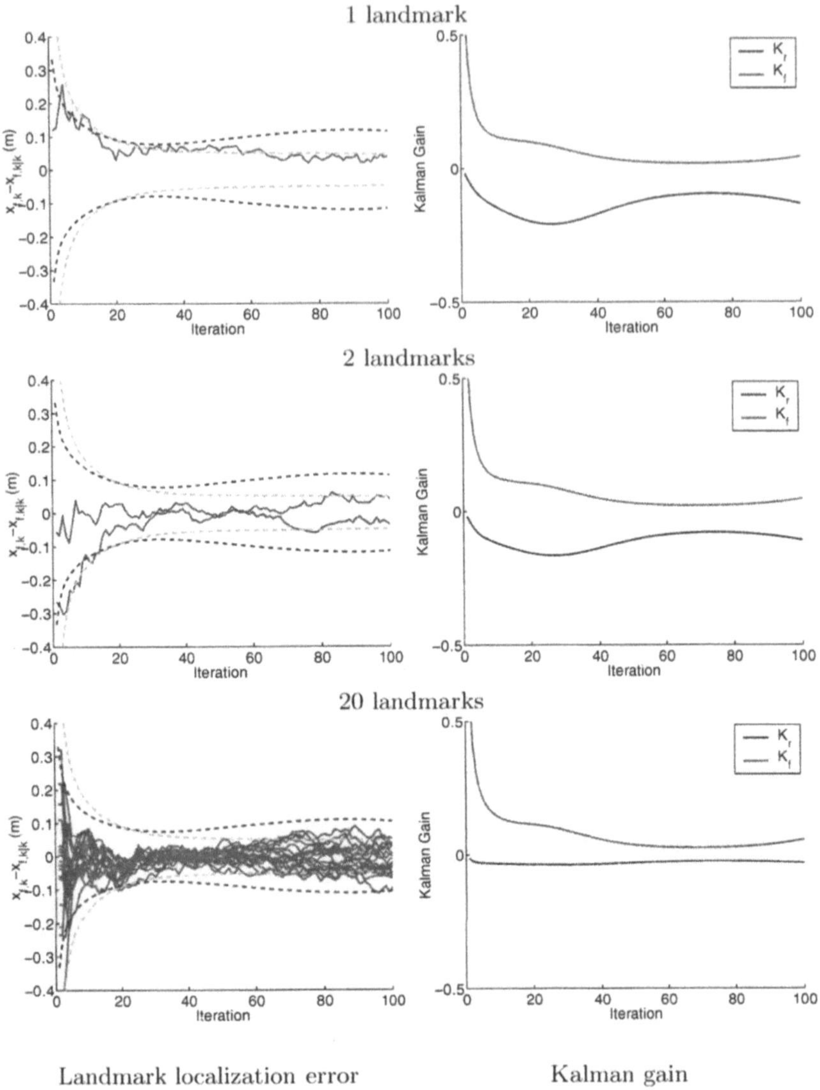

Fig. 2.18. Full-covariance fully observable SLAM with artificial process noise for a monobot in a sinusoidal path starting at $\mathbf{x}_{r,0|0} = -1m$, and with 100 iterations. A fixed external sensor is used for the measurement of the vehicle pose, and a pseudo-noise covariance ($\mathbf{V}_f = 0.05^2\mathbf{I}$) is used to prevent the landmark Kalman gains from converging to zero.

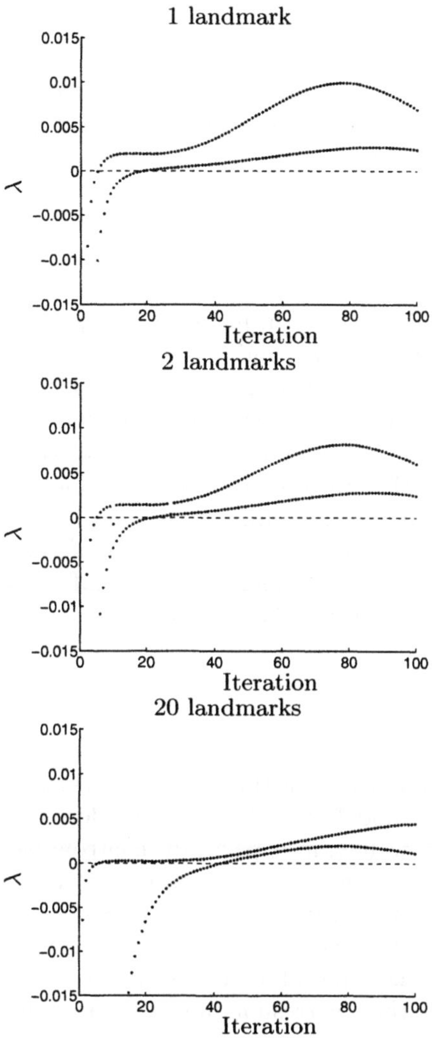

Fig. 2.19. The Cramer Rao lower bound holds only when the eigenvalues of $(\mathbf{P}_{k|k} - \mathbf{J}^{-1})$ are positive.

2.9 Bibliographical Notes

Gibbens *et al.* [41] give a solution to the 1-D Brownian motion SLAM case. In their solution, the state error covariance is linked to the total number of landmarks in the form of the total Fisher information I_T.

$$I_T = \sum_{1}^{n} \frac{1}{\sigma_w^2} \tag{2.67}$$

The expression indicates the "*informational equivalence of the measurements and the innovations*" [18], and was derived from a simpler likelihood function than the one in Equation 2.2; one that does not contain the fully correlated characteristics of the measurement model.

The issues of controllability and observability in Kalman filter design are covered in the book by Bar-Shalom *et al.* [18]. Southall *et al.* address the issue in the context of tracking for an autonomous agricultural application [80].

A wonderful text on estimation theory is the aforementioned book by Bar-Shalom. The Cramer Rao lower bound and the information matrix are also discussed in an article by Dowski [32].

2.10 Conclusions

The fundamental contributions of this chapter is that we show how the full-correlation of the map model in EKF-SLAM hinders full observability of the state estimate, and how having a partially controllable state model makes the filter to stop working after a few iterations. Partial observability makes the final map dependant on the initial observations, and does not guarantee convergence to a positive definite covariance matrix. Partial controllability on the other hand, makes the filter believe it has accurate estimates of the landmark states, with their corresponding Kalman gains converging rapidly to zero. That is, after a few steps, innovations are useless for the refinement of landmark locations, and contribute only to the revision of the vehicle pose.

Partial observability can easily be solved by anchoring the map to the first landmark observed, or by having an external sensor that sees the vehicle at all times. Partial controllability on the other hand, can be only palliated to some extent by adding artificial process noise to the landmark estimates.

Additionally, the divergence caused by nonlinearities in SLAM can be reduced either by assuming larger uncertainty of the vehicle model (augmenting the process noise covariance), and with the online computation of measurement noise covariance matrices.

3

Suboptimal Filter Stability

When a stochastic system is partially controllable, such as in the case of SLAM, the Gaussian noise sources $\mathbf{v}_k$ do not affect all of the elements of the state space. The diagonal elements of $\mathbf{P}$ corresponding to these incorruptible states will be driven to zero by the Kalman filter, and once this happens, these estimates will remain fixed and no further observations will alter their values. The dynamics of the model assume the landmarks are fixed elements, for which no process noise is considered. Therefore, their associated noise covariance (its determinant) will asymptotically tend to zero [31]. The filter gain for the landmark states will also tend to zero. Figure 3.1 shows two new simulations for a linear SLAM case, a monobot under Brownian motion with one and two landmarks. The simulations show the evolution of the localization errors for both the monobot and the landmarks, and the reduction of the landmark part of the Kalman gain, due to the uncontrollability of the system. The only way to remedy this situation is to add a positive definite pseudo-noise covariance to those incorruptible states [18].

In Figure 3.1, the vehicle location is indicated by the darkest curve at the $-1m$ level in the first row of plots. In the same set of plots, and close to it, is a lighter curve indicating the vehicle location estimate as computed by the filter, along with 2σ bounds on such estimate shown as dotted lines. The dark straight lines at the $1m$ level indicate the landmark location estimates; and the lighter curves are noise corrupted signals of sensor measurements. Also shown, are a pair of dotted lines for 2σ bounds on the landmark location estimates. The second row of plots shows the vehicle location error only, and its corresponding variance, also on the form of 2σ dotted bounds. See how the localization error has non-zero mean due to partial observability, an undesirable feature in Kalman filtering. The third row shows non-zero mean landmark state estimate errors. And, the last row shows the Kalman filter gains both for the vehicle and landmark revision terms. The Kalman gains for the revision of the landmark estimates rapidly tend to zero, the reason being that these states are uncontrollable.

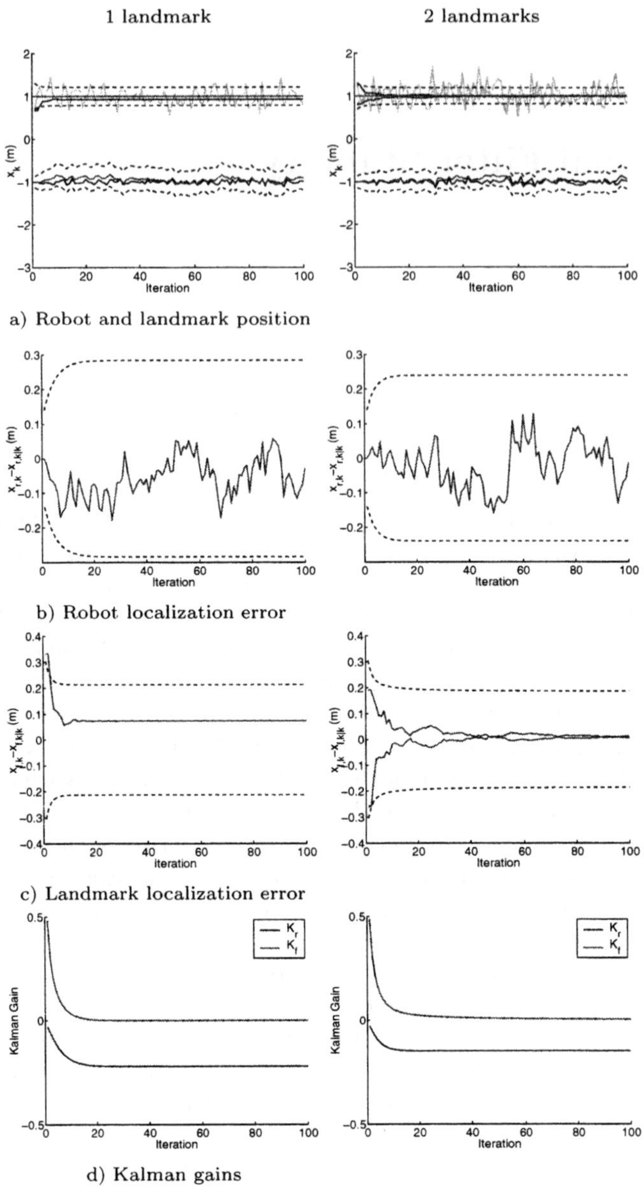

Fig. 3.1. Partially observable SLAM for a monobot during Brownian motion with 100 iterations (see text).

3.1 $O(n)$ but Unstable Partially Observable SLAM

One way to add pseudo-noise to the model is by diagonalizing the state error covariance matrix [44, 45, 51]. This technique is often used to reduce the time complexity of the algorithm from $O(n^2)$ to $O(n)$. The result is a suboptimal filter that will compute inflated estimates for the vehicle and landmark covariances, that has the computational advantage of being uncorrelated. The addition of a covariance term $\Delta\mathbf{P}$ to the a priori state covariance estimate

$$\mathbf{P}_{k+1|k} = \mathbf{F}\mathbf{P}_{k|k}\mathbf{F}^{\top} + \mathbf{G}\mathbf{Q}\mathbf{G}^{\top} + \Delta\mathbf{P} \tag{3.1}$$

is equivalent to providing a new form to the plant noise Jacobian $\mathbf{G}' = \begin{bmatrix}\mathbf{G}\ \mathbf{I}\end{bmatrix}$

$$\mathbf{P}_{k+1|k} = \mathbf{F}\mathbf{P}_{k|k}\mathbf{F}^{\top} + \mathbf{G}'\begin{bmatrix}\mathbf{Q} & \\ & \Delta\mathbf{P}\end{bmatrix}\mathbf{G}'^{\top} \tag{3.2}$$

$\Delta\mathbf{P}$ may be chosen, for example, such as to minimize the trace of a resulting block diagonal $\mathbf{P}$ in (3.1) [51].

Choosing a full rank $\Delta\mathbf{P}$ is equivalent to having noise input to more states than those that can be observed with the filter. In this case, because of partial observability, both vehicle and landmark variance estimates become unbounded. Figure 3.2 shows this for the same monobot experiment as in the previous simulation. This phenomena was first observed in [51] using relative maps.

Not only both the vehicle and landmark state estimation variances become unbounded. Also, thanks to the full controllability of the system, the Kalman gain for the revision of the landmark states is greater than zero; but still, does not converge to a steady state value. We believe that the addition of pseudo-noise should be performed only at most, in the amount of states equal to the dimension of the observable subspace.

3.2 $O(n)$ and Stable Partially Observable SLAM

One solution to the problem of instability during covariance inflation, is to decorrelate only the landmark state estimates, and to preserve all vehicle to landmark correlations [88].

$$\Delta\mathbf{P} = \begin{bmatrix}\mathbf{0} & \\ & \mathbf{Q}_f\end{bmatrix} \tag{3.3}$$

such that $\mathbf{P}_f + \mathbf{Q}_f$, the map part of the state error covariance, is block diagonal.

Figure 3.3 shows a partially observable monobot under Brownian motion for which only the landmark part of the state error covariance matrix has been decorrelated. The algorithm does converge to a steady state solution under this circumstances, and still can be implemented in real time. The one landmark case is identical than the original case, since a linear one landmark map is already diagonal (scalar actually).

For the two-landmark case, the landmark variance estimate is greater than the optimal solution shown in the third column in Figure 3.1 since the covariance has been inflated during decorrelation. Furthermore, now that the system is controllable, the Kalman gains for the landmark state estimates do not become zero, and they converge to a steady state value.

Moreover, we can see experimentally, that the covariance inflation suboptimal partially observable SLAM converges only when

$$\text{rank } \Delta\mathbf{P} \leq \text{rank } \mathcal{O} \tag{3.4}$$

3.3 $O(n)$ and Stable Fully Observable SLAM

Consider now the fully observable case from the previous Section. If we add pseudo-noise to the vehicle as well as to the landmark states, the covariance will reach a steady-state value, and the Kalman gain will not be zero, at least, in the linear case. Figure 3.4 shows this results diagonalizing the whole state error covariance (not only the landmark part of $\mathbf{P}$).

In this latter experiment, the state error variances reach lower values than those in the partially observable case. The solution of the Riccati equation is now independent of the initial covariance estimate $\mathbf{P}_{0|0}$.

We have observed experimentally however, that with a nonlinear vehicle model, it is best to also decorrelate only the map part of the state error covariance, even in the fully observable case.

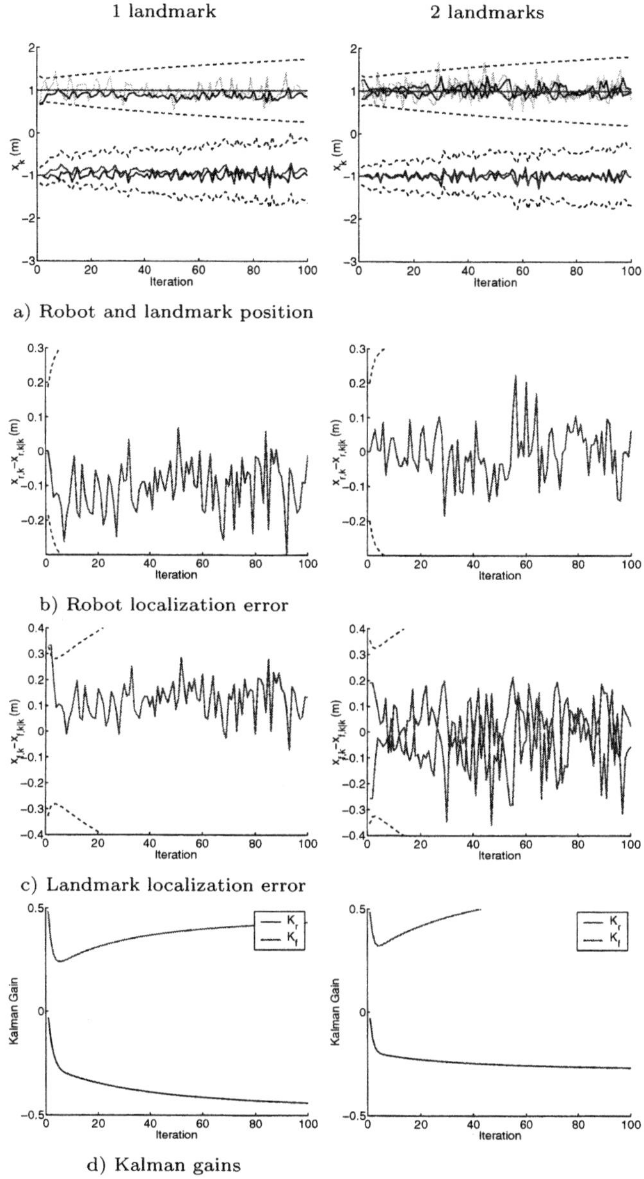

Fig. 3.2. Partially observable SLAM for a Brownian motion monobot with 100 iterations. The entire state error covariance is decorrelated with the minimal trace solution [51]. By decorrelating the entire state error covariance matrix, the covariance estimates become unbounded.

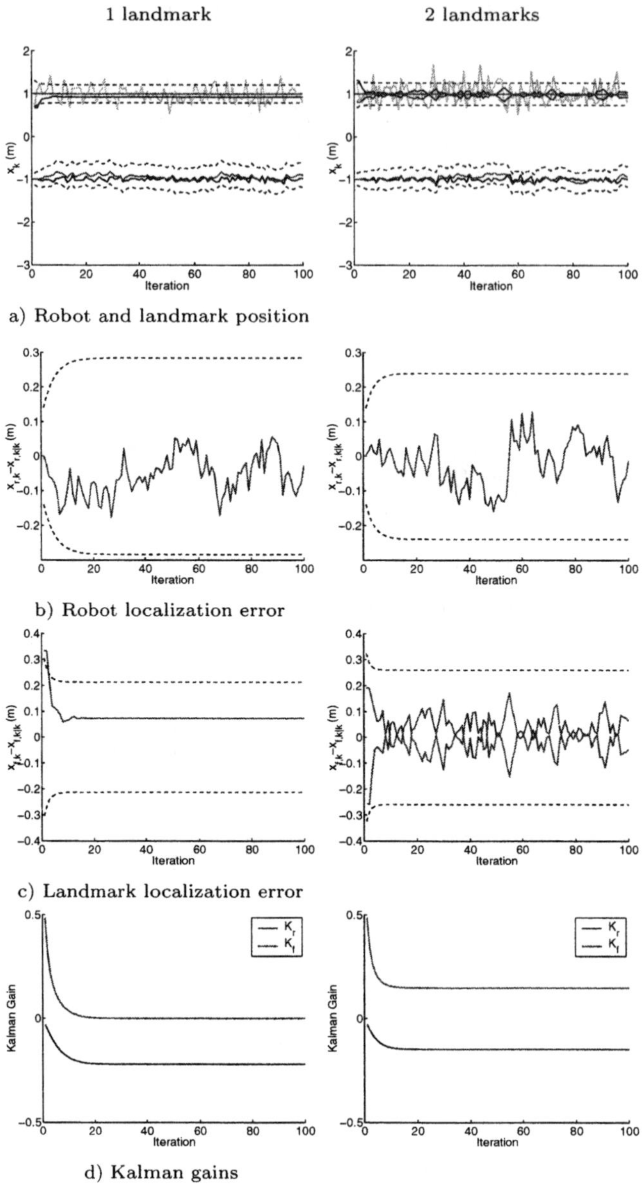

Fig. 3.3. Partially observable SLAM for a Brownian motion monobot with 100 iterations. The state error covariance is decorrelated only for the landmark part of the state vector, with the minimal trace solution. By decorrelating only the map part of the state error covariance matrix, we preserve filter stability.

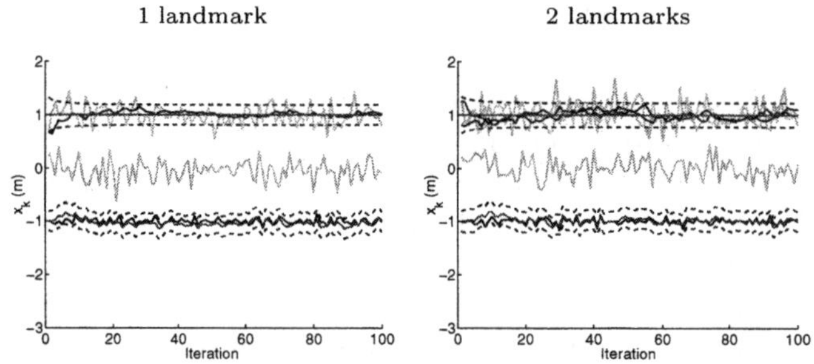

a) Robot and landmark localization

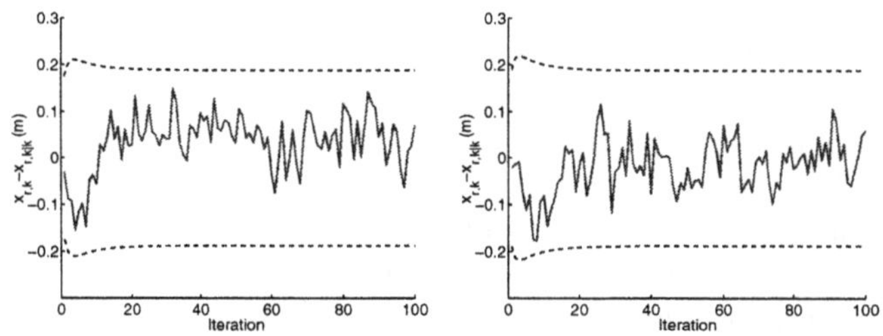

b) Vehicle error

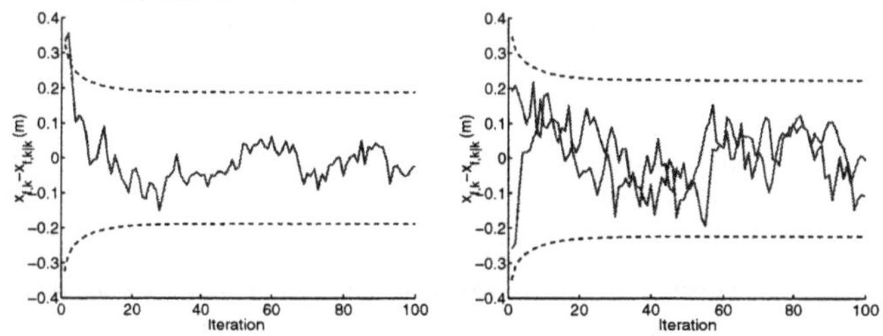

c) Landmark localization error

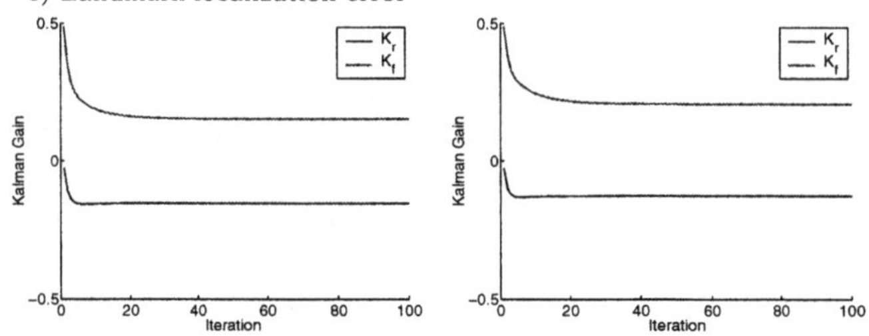

d) Kalman gains

Fig. 3.4. Fully observable SLAM for a Brownian motion monobot with 100 iterations. The entire state error covariance is decorrelated with the minimal trace solution. In the linear case, it is possible to decorrelate the entire state error covariance matrix, and still preserve filter stability.

3.4 Experimental Results

We show now results on a series of experiments for a nonlinear vehicle with an also nonlinear measurement model, using the ACFR - University of Sydney database [69]. The test run used corresponds to a car-like vehicle at the University Car Park. The landmarks used are tree trunks, as measured with a laser range finder. The reconstructed maps are compared to GPS ground truth for accuracy. The first experiment corresponds to a typical partially observable SLAM run, in which the entire state error covariance is being decorrelated as discussed in Section 3.1. Figure 3.5 plots results on this run, showing in rows b) and d) unbounded covariances both for the vehicle and landmark state estimates, due to the naïve covariance inflation method used.

The second experiment corresponds to the same partially observable SLAM conditions, but decorrelating only the map part of the state error covariance. Adding pseudo-noise to the landmark states during the inflation procedure amounts to making the system controllable; and doing so for as many states as those observable, produces both vehicle and landmark bounded state covariance estimates. This is shown in Figure 3.6, frames b) and d). Figure 3.8 frame a) shows the actual vehicle path and landmark location estimates recovered by the algorithm, compared to GPS ground truth for the beacons. Note that even when the "relative" map is consistent [31], it is slightly rotated and shifted from the actual beacon locations. The amount of this shift depends on the initial vehicle uncertainty, i.e., the initial filter conditions, and can be seen in Figure 3.6, frame c).

The last experiment shown corresponds to a fully observable SLAM run (using the first observed beacon as an anchor), and also decorrelating only the map part of the state error covariance. In this case, the vehicle and landmark covariance estimates do not depend on the initial filter conditions, and thus are significantly reduced. This is shown in frames b) and d) in Figure 3.7. The absolute landmark estimate error is also significantly reduced, as shown in Figure 3.7, frame c). Figure 3.8 frame b) shows the actual vehicle path and landmark estimates as recovered by the filter. The beacon shown in the center of the plot is used as an anchor to the map, and no state estimate is computed for it. This last map was obtained with a suboptimal linear-time SLAM algorithm that has both bounded covariance estimates, and independence on the filter initial conditions; thus producing a fast and accurate absolute map.

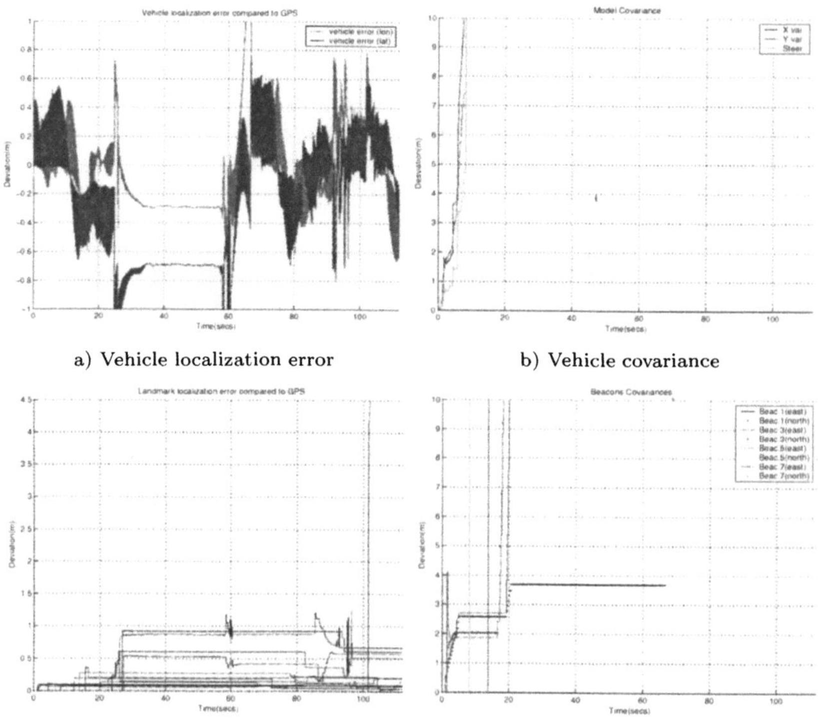

a) Vehicle localization error

b) Vehicle covariance

c) Landmark localization errors

d) Landmark covariance

Fig. 3.5. Partially observable SLAM for a car-like vehicle at the University of Sydney Car Park. The entire state error covariance matrix is decorrelated with the minimal trace solution [51].

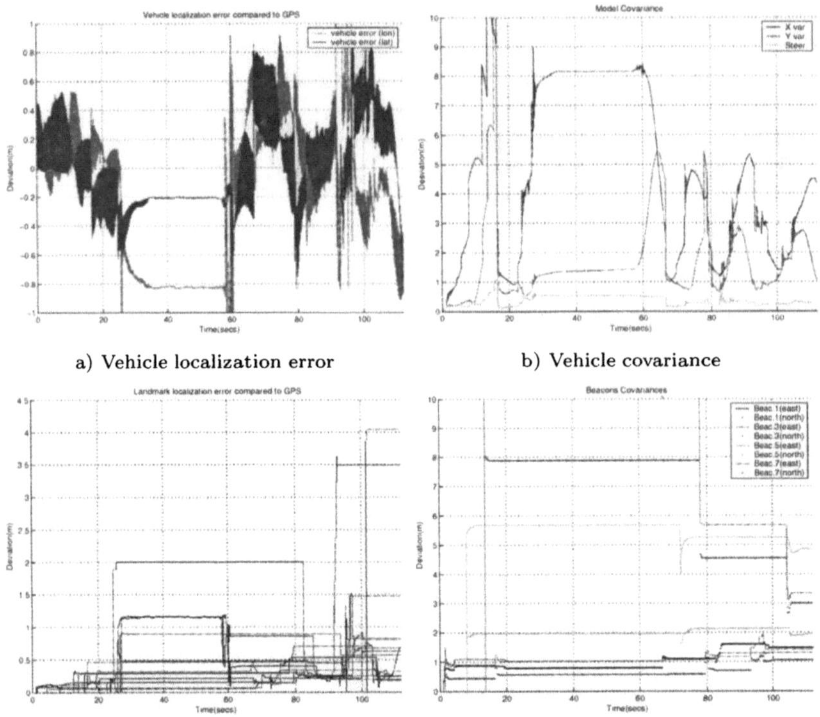

a) Vehicle localization error

b) Vehicle covariance

c) Landmark localization errors

d) Landmark covariance

Fig. 3.6. Partially observable SLAM for a car-like vehicle at the University of Sydney Car Park. Only the map part of the state error covariance matrix is decorrelated with the minimal trace solution. By adding controllability to as many states as those that are observable, the filter remains stable, and the estimated covariances remain bounded.

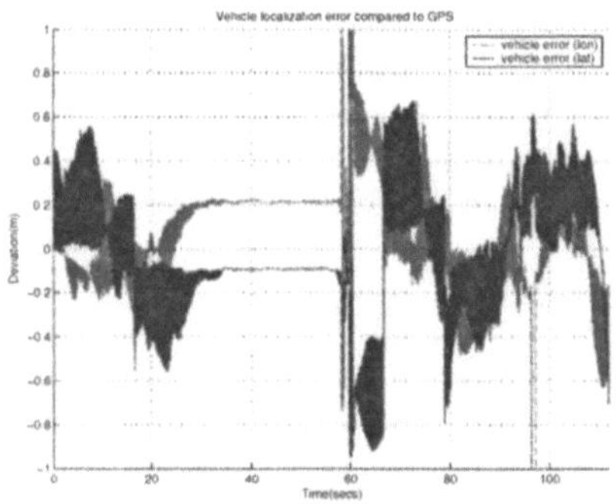

a) Vehicle localization error

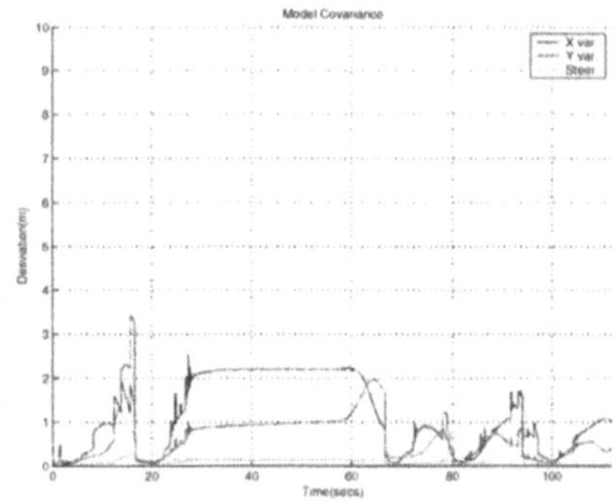

b) Vehicle covariance

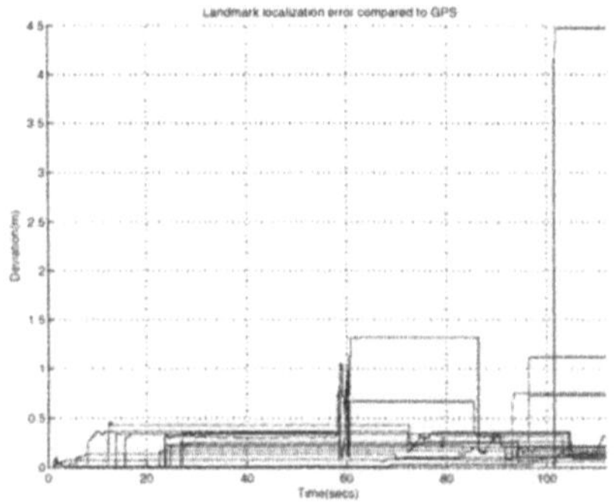

c) Landmark localization errors

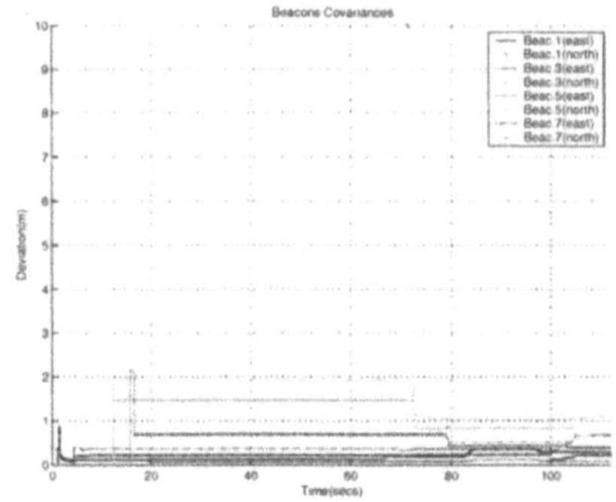

d) Landmark covariance

Fig. 3.7. Fully observable SLAM for a car-like vehicle at the University of Sydney Car Park. Only the map part of the state error covariance is decorrelated with the minimal trace solution. Full observability guarantees independence of the filter initial conditions, and an accurate absolute map is obtained, with smaller covariance estimates than its relative counterpart.

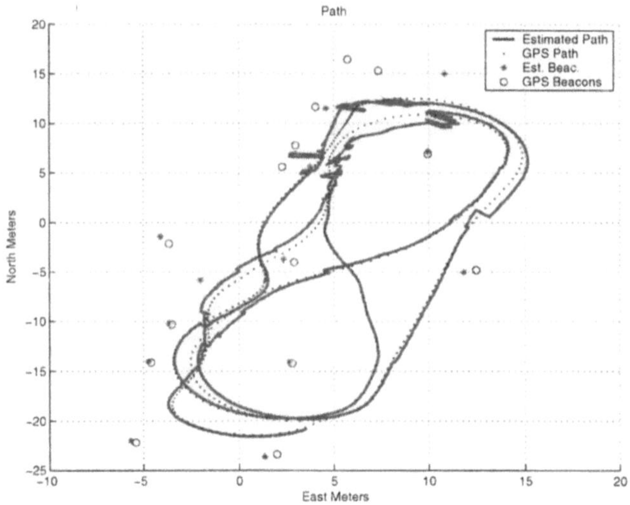

a)

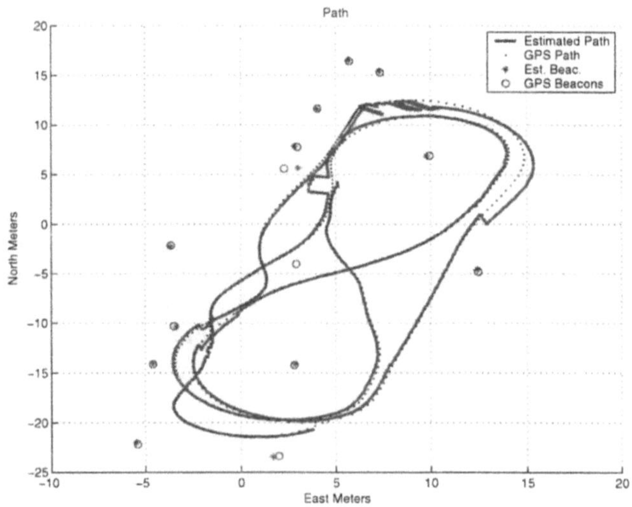

b)

Fig. 3.8. Vehicle path and landmark location estimates, compared to GPS ground truth for an a) partially observable suboptimal SLAM run, and a b) fully observable suboptimal SLAM run; both with decorrelation of only the map part of the state error covariance matrix.

4

Unscented Transformation of Vehicle States

We have already seen that the Extended Kalman Filter (EKF) is the most widely accepted tool for solving SLAM [31, 85]. One drawback however with the use of the EKF, is in the linear propagation of means and covariances. Vehicle and sensor models are usually of a very high nonlinear nature, and the effects of linearization required in the EKF can lead to filter divergence [53].

This situation has prompted the use of particle filters for a non parametric approximation of vehicle and map probability density functions in SLAM. Particle filters approximate the state space by random sampling the posterior distribution, and may require many samples to accurately model the nonlinear effects in both vehicle and measurement models. A middle ground is to use a deterministic approach for the nonlinear propagation of means and covariances. One such solution is the use of the Unscented Kalman Filter (UKF) [50, 55]. An unscented transformation is similar to a particle filter in that it samples the *pdf*, but instead of doing it randomly, a careful selection of deterministic *sigma* points is made so as to preserve the moments of the distribution.

Deterministically choosing the particles is a computationally efficient solution for the nonlinear propagation of means and covariances, but doing so for the full state vector in SLAM may not be appropriate. There is no need to use particles in the computation of the map prior, given its linear nature. Thus, by using the Unscented Transformation (UT) only for the vehicle states we are able to reduce the computational complexity (compared to a full UKF), and to produce, at the same time, tighter covariance estimates.

The remaining of this chapter is structured as follows. In Section 4.1, the UKF is explained, detailing the consequences of nonlinearly propagating the entire vehicle and map state vector, as opposed to only propagating the vehicle states. Section 4.3 is devoted to a numerical comparison of the three approaches: EKF, UKF, and vehicle only UT. Finally, Section 4.4 contains some concluding remarks.

4.1 Nonlinear Propagation of State Estimates

Julier and Uhlmann [48, 52, 55] introduced a filter called the Unscented Kalman Filter (UKF) founded on the intuition that it is easier to approximate a Gaussian distribution than to approximate nonlinear functions. The UKF not only leads to more accurate results than the EKF, but also, it generates better estimates of the state covariances. The reason, the UKF approximates the posterior mean and covariance accurately to the 2nd order for any nonlinearity, in contrast with the EKF which only takes into account the first order term of a Taylor series during linearization.

The Unscented Transform (UT) forms the core of the UKF algorithm, it consists in choosing a set $\mathcal{S}$ of points (*sigma* points $\mathcal{X}^i$ and weights W^i) so that their mean and covariance are $\mathbf{x}$ and $\mathbf{P}$. The nonlinear model in (1.4) is applied to each point, and the weighted statistics of the transformed points form an estimate of the nonlinearly transformed mean and covariance.

The UKF algorithm is similar in structure to the EKF algorithm. Once the *sigma* points are obtained (see Appendix C), the a priori estimates are evaluated with

$$\mathcal{X}^i_{k+1|k} = \mathbf{f}(\mathcal{X}^i_{k|k}, \mathbf{u}_k, \mathcal{V}^i_k) \tag{4.1}$$

$$\mathbf{x}_{k+1|k} = \sum_{i=0}^{p} W^i \mathcal{X}^i_{k+1|k} \tag{4.2}$$

$$\mathbf{P}_{k+1|k} = \sum_{i=0}^{p} W^i (\mathcal{X}^i_{k+1|k} - \mathbf{x}_{k+1|k})(\mathcal{X}^i_{k+1|k} - \mathbf{x}_{k+1|k})^\top \tag{4.3}$$

In the UKF the innovations are

$$\mathcal{Z}^i_{k+1|k} = \mathbf{h}(\mathcal{X}^i_{k+1|k}) + \mathcal{W}^i_k \tag{4.4}$$

$$\mathbf{z}_{k+1|k} = \sum_{i=0}^{p} W^i \mathcal{Z}^i_{k+1|k} \tag{4.5}$$

$$\mathbf{P}^{zz}_{k+1|k} = \sum_{i=0}^{p} W^i (\mathcal{Z}^i_{k+1|k} - \mathbf{z}_{k+1|k})(\mathcal{Z}^i_{k+1|k} - \mathbf{z}_{k+1|k})^\top \tag{4.6}$$

And the a posteriori estimates are

$$\mathbf{P}^{xz}_{k+1|k} = \sum_{i=0}^{p} W^i (\mathcal{X}^i_{k+1|k} - \mathbf{x}_{k+1|k})(\mathcal{Z}^i_{k+1|k} - \mathbf{z}_{k+1|k})^\top \tag{4.7}$$

$$\mathbf{K} = \mathbf{P}^{xz}_{k+1|k}(\mathbf{P}^{zz})^{-1} \tag{4.8}$$

$$\mathbf{x}_{k+1|k+1} = \mathbf{x}_{k+1|k} + \mathbf{K}\tilde{\mathbf{z}}_{k+1|k} \tag{4.9}$$

$$\mathbf{P}_{k+1|k+1} = \mathbf{P}_{k+1|k} - \mathbf{K}\mathbf{P}^{zz}\mathbf{K}^\top \tag{4.10}$$

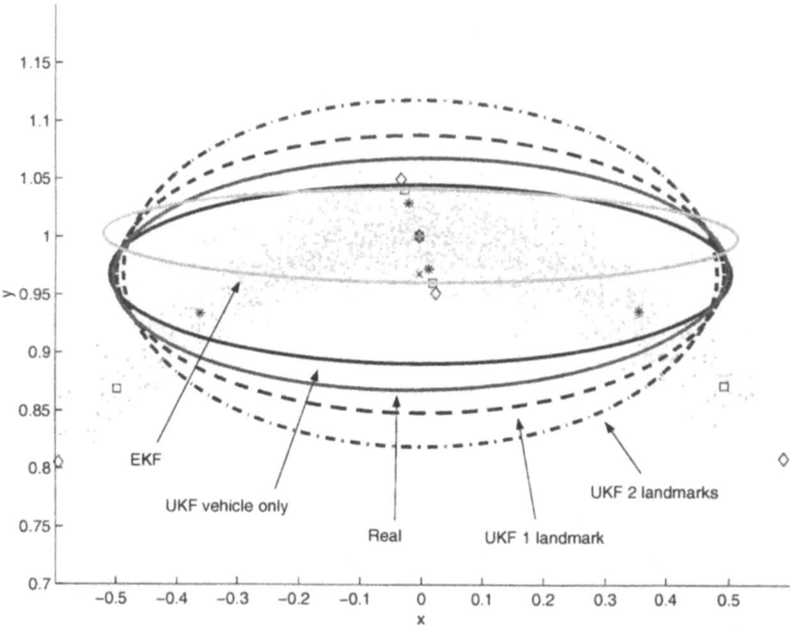

Fig. 4.1. Unscented transformation of the augmented state vector produces more σ points that project to the mean. To preserve the statistics, the remaining σ points are pushed away from the mean, thus producing more conservative estimates of the covariance in the nonlinear case. In red (x): actual mean and covariance; in green (+): linear transformation of mean and covariance; in blue (*): nonlinear transformation of mean and covariance for robot only; in blue dashed (□): nonlinear transformation for robot and one landmark; in blue dash dot (◇): nonlinear transformation for robot and 2 landmarks.

The terms $\mathcal{V}^i$ and $\mathcal{W}^i$ are *sigma* points for the noise terms. To compute them, an augmented point set is built from $\mathbf{x}, \mathbf{P}, \mathbf{Q}$ and $\mathbf{R}$. See [55] for details.

The update estimation in the UKF algorithm requires an augmented *sigma* point set for the entire state vector in (4.7). The computation of these points requires a Cholesky factorization of $\mathbf{P}$, with computational complexity $O(n^3)$.

4.2 UT of Vehicle States

In SLAM, the dimension of the state vector is proportional to the number of landmarks in the map, and every time a new feature is added to the map, the state vector is also augmented. Moreover, when the UKF is used, the required number of *sigma* points is also proportional to the number of landmarks in the map. These *sigma* points are symmetrically distributed along the hyperellipsoid representing the covariance of the entire state space. Now, every time a new landmark is added, the corresponding *sigma* points will map the new state space directions, and project to the mean on the vehicle space hyperplane, with the rest of the points being scaled accordingly, in order to preserve the first and second order statistics of the entire distribution, i.e., $E[\tilde{\mathbf{x}}]$ and $E[\tilde{\mathbf{x}}\tilde{\mathbf{x}}^{\top}]$.

Looking at the projected hyperellipsoid representing the vehicle covariance we see that as the number of landmarks increases, the UT maintains the true mean of the vehicle prior, but it underestimates the vehicle covariance. Figure 1 shows this situation. In the plot, a typical nonlinear motion model of one meter with translational variance of $2cm$ and rotational variance of $15°$ is sampled 1000 times (dots). The true mean and 2σ hyperellipsoid for the true covariance are also plotted (cross mark and continuous line). The linear transformation of the original *pdf* (the prior step of the EKF) is the smallest of the hyperellipsoids shown. In the plot, the mean computed by the EKF is slightly above the true mean, and the linear transformation of the vehicle covariance is largely overestimated. Now, the UT for this no-landmark model is at the true mean, with also an overestimated covariance, but not as much as the linear transformation (the mean and corresponding 5 *sigma* points are plotted as stars, and the 2σ covariance hyperellipsoid is plotted as a continuous line. Adding one two-dimensional landmark to the map at (0,0) and with variance $1cm$, the required number of *sigma* points increases from 5 to 9, and their projection onto the Cartesian vehicle coordinates is shown with small boxes. Notice how the new *sigma* point locations emulate the previous set, but are pushed away from the mean (scaled), with more points located precisely at the mean. The corresponding covariance hyperellipsoid projection (dashed) is slightly underestimated with respect to the previous UT and the original covariance. Finally, adding one more landmark to the map, the required number of *sigma* points is 13, and their projection onto the vehicle state space is shown as diamonds, with the corresponding 2σ covariance hyperellipsoid shown in dash dot. Even when the computed mean is equal to the one computed in the previous two cases, the *sigma* points are further pushed away from the mean, and the corresponding covariance is also underestimated. The conclusion is that when the number of landmarks increases significantly, the UT is still good at computing the nonlinear mean estimate for the vehicle position, but it largely underestimates its covariance.

The variance estimates of the dynamic states (a priori vehicle location estimation) should not depend on the number of static states on the model (landmarks), but only on the characteristics of the motion model. The map entries being static have a linear model (identity in fact), and no UT is necessary for this transformation. The underestimation of the covariance is the result of projecting a higher dimensional hyperellipsoid of such transformation onto the vehicle states hyperplane. For this reason we propose to apply the UT only to the vehicle states, and not to the full state vector during the prediction step.

The plant Jacobian matrices in (1.16) can be decomposed into two block diagonal matrices, explicitly differentiating vehicle derivatives $\mathbf{F}_r$ and $\mathbf{G}_r$. Equation (1.16) can be rewritten as

$$\mathbf{P}_{k+1|k} = \begin{bmatrix} \mathbf{F}_r\mathbf{P}_{r,k|k}\mathbf{F}_r^\top + \mathbf{G}_r\mathbf{Q}\mathbf{G}_r^\top & \mathbf{F}_r\mathbf{P}_{rf,k|k} \\ (\mathbf{F}_r\mathbf{P}_{rf,k|k})^\top & \mathbf{P}_{f,k|k} \end{bmatrix} \tag{4.11}$$

Our algorithm substitutes the a priori computation of the vehicle covariance in the EKF, with the one computed using the UT; while preserving the rest of the covariance matrix. That is, the upper left submatrix is substituted with

$$\begin{aligned} \mathbf{P}^*_{r,k+1|k} &= \mathbf{P}_{r,k+1|k} + \Delta \\ &= \sum_{i=0}^{p} W^i(\mathcal{X}^i_{r,k+1|k} - \mathbf{x}_{r,k+1|k})(\mathcal{X}^i_{r,k+1|k} - \mathbf{x}_{r,k+1|k})^\top \end{aligned} \tag{4.12}$$

We have seen empirically that by substituting the submatrix $\mathbf{P}_{r,k+1|k}$ with $\mathbf{P}^*_{r,k+1|k}$ in (4.11), the *psd* properties of $\mathbf{P}^*_{k+1|k}$ still hold. That is, by underestimating $\mathbf{P}^*_r$ with the UT, and still computing the rest of Jacobians in $\mathbf{P}^*_{k+1|k}$ with the linear EKF, the condition

$$\mathbf{P}^*_{k+1|k} = \begin{bmatrix} \mathbf{P}_{r,k+1|k} + \Delta & \mathbf{P}_{rf,k+1|k} \\ \mathbf{P}^\top_{rf,k+1|k} & \mathbf{P}_{f,k+1|k} \end{bmatrix} \geq 0 \tag{4.13}$$

still holds.

4.3 Experimental Results. EKF, UKF, and Vehicle-Only UT

We use the University of Sydney Car Park dataset [69] to show estimation results when the prediction is made with the nonlinear approach proposed in this article, and compare them with both an EKF, and an UKF. With the proposed technique, we have been able to reduce the bias caused by linearization of the nonlinear plant model in the EKF.

The nonlinear vehicle and measurement models used in our experiments are shown in Section 1.2, and resemble to some extent those in [44]. The only substantial difference is in the use of an anchor feature not under estimation in the measurement model, in order to guarantees full observability [10, 89].

Figures 4.2-4.4 contain plots comparing the three algorithms, for the vehicle location error, the vehicle location variance, and the final map, respectively. The first column of plots in Figure 4.2 corresponds to a fully observable Extended Kalman Filter, in which the first observed landmark (located at coordinates $(2.8953, -4.0353)$) is used as a global anchor. The second column of plots corresponds to a full implementation of the Unscented Kalman Filter.

The last column of plots corresponds to our hybrid implementation: using the Unscented Transformation for a nonlinear approximation of vehicle priors, and filling the rest of the a priori covariance estimate with the required vehicle to landmark Jacobians, plus the typical a posteriori computations in the EKF. This technique is not only computationally more efficient (it requires on the one hand a small and fixed number of *sigma* points in the computation of the UT, 7 in our case, and also, it can make efficient use of sequential innovation as in the traditional EKF approach to SLAM), but reduces significantly the errors made during linearization in the computation of the vehicle priors.

In congruence with our conclusions from Section 4.1, the Unscented Kalman Filter, which utilizes the entire state vector in the computation of the *sigma* point set, ends up underestimating the vehicle localization covariance at the beginning of the experiment. This can be appreciated in Figure 4.2e. However, once observations are made, the algorithm makes a good job at reducing the overall vehicle covariances. The vehicle location error estimate is also slightly smaller for the entire UKF approach, compared to the other two algorithms. See frame b in Figure 4.2.

Unfortunately, the UKF is not only computationally more expensive (it requires $2n + 1$ *sigma* points at each iteration, although there are other approaches that require only $n + 2$ *sigma* points [50]), but it might end up overestimating the final vehicle and map covariances, producing data association errors in the long run. This can be appreciated in the error peaks near $t = 90$ secs. in Figures 4.2a-b, and in the lower part in Figure 4.3b. Nearest Neighbor χ^2 compatibility tests are used for data association. We believe that the overestimation of covariances is what makes this test to fail when using the full UKF.

By computing the vehicle priors using the Unscented Transformation, the estimated vehicle localization error is smaller during the entire run than the error computed using the EKF; and the computed covariances are quite similar for both algorithms. However, there is a significant advantage at using the UT vs. the EKF: when there are no measurements present, the nonlinear transformation of the vehicle estimates makes a good job at keeping the vehicle location closer to the desired vehicle path. This can be seen around coordinates $(15, 5)$ in the three plots pertaining the final vehicle path in Figures 4.3 and 4.3.

4.4 Conclusion

The Unscented Transformation allows a better nonlinear mean and variance estimation than the Extended Kalman Filter. There is no need however in using the Unscented Transformation for the entire vehicle-map state, given the linearity in the map part of the model.

Large underestimation errors in the calculation of the covariance priors can be made when the full state vector is used in the computation of the *sigma* point set for the Unscented Transformation in SLAM. This is because the more landmarks are added to the map, the larger the number of *sigma* points in the set that get projected to the vehicle mean in the vehicle localization hyperplane; and the remaining *sigma* points need to be scaled up to preserve both the mean and covariance.

By applying the Unscented Transformation only to the vehicle states we get more accurate covariance estimates, and a more computationally efficient nonlinear transformation of the means and variances in SLAM. In the presented approach, the a posteriori state estimation is made using a fully observable EKF step, thus preserving the same computational complexity as the EKF with sequential innovation.

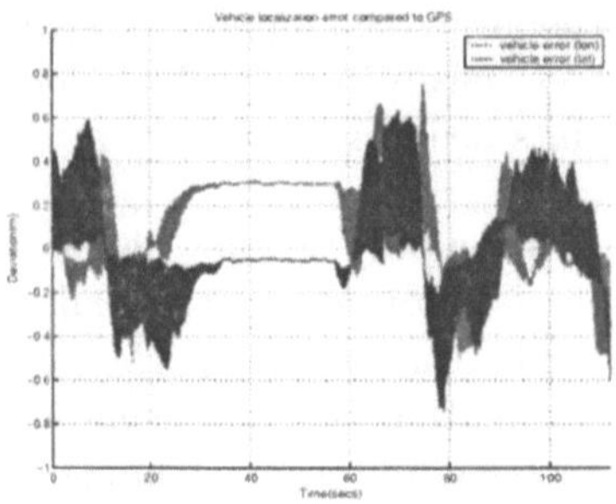

(a) EKF Vehicle error

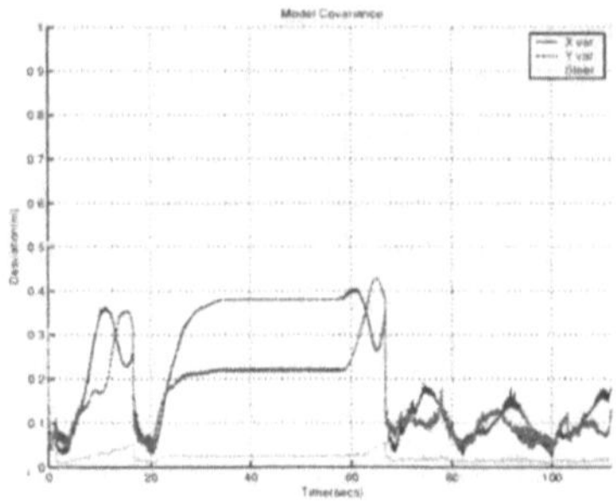

(b) EKF Vehicle covariance

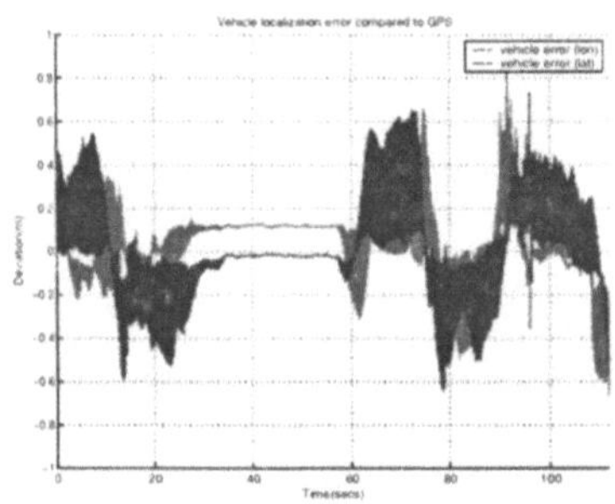

(c) UKF Vehicle error

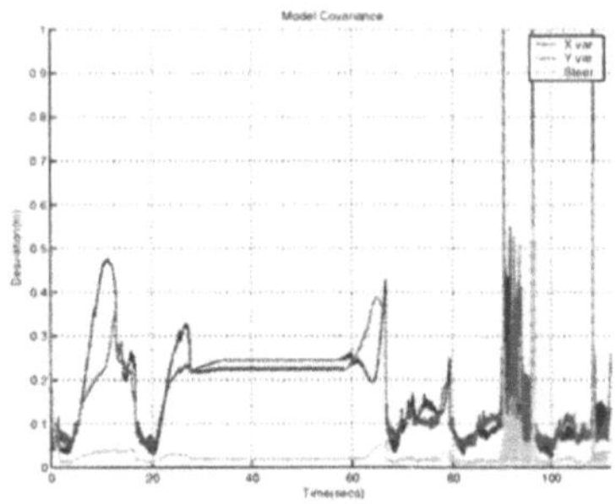

(d) UKF Vehicle covariance

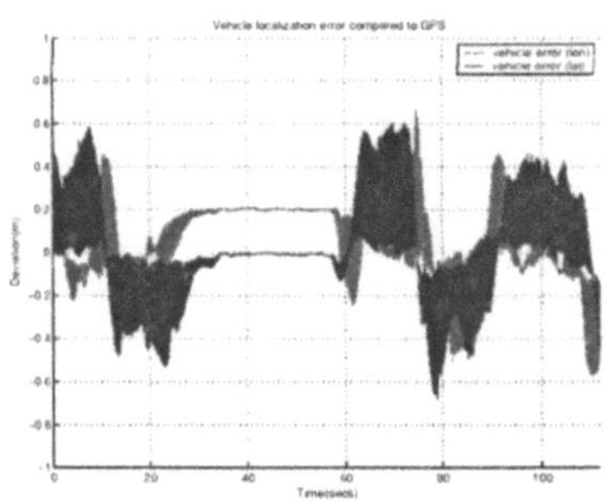

(e) UT Vehicle error

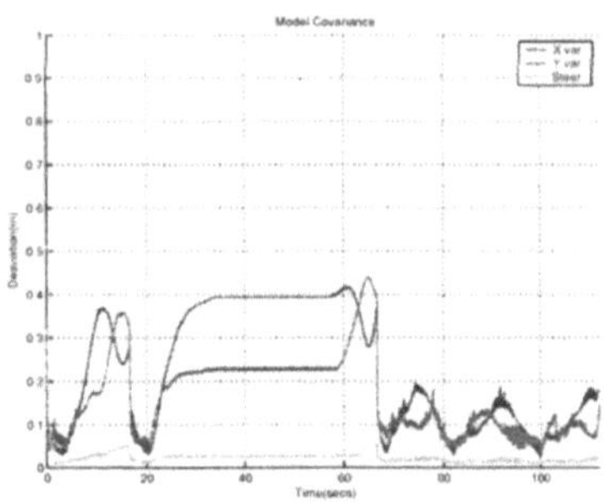

(f) UT Vehicle covariance

Fig. 4.2. Comparison of the Extended Kalman Filter, the Unscented Kalman Filter, and Unscented Transformation of Vehicle States only on the Car Park dataset from the University of Sydney. The Unscented Transformation of Vehicle states is not only computationally more efficient than the full UKF, but avoids data association errors due to covariance overestimation in the full UKF.

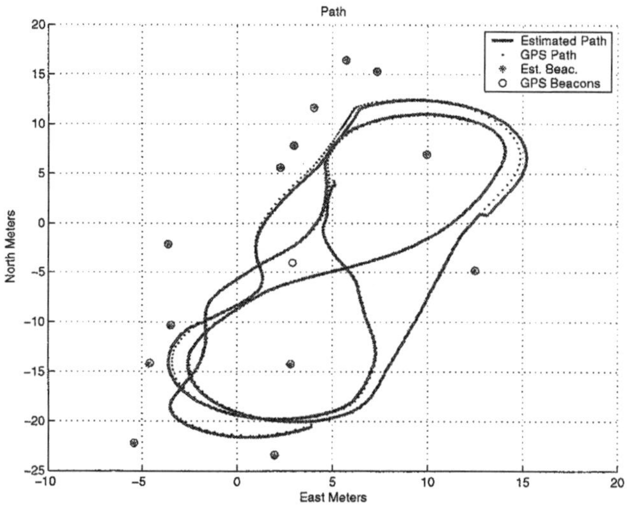

(a) EKF Final map

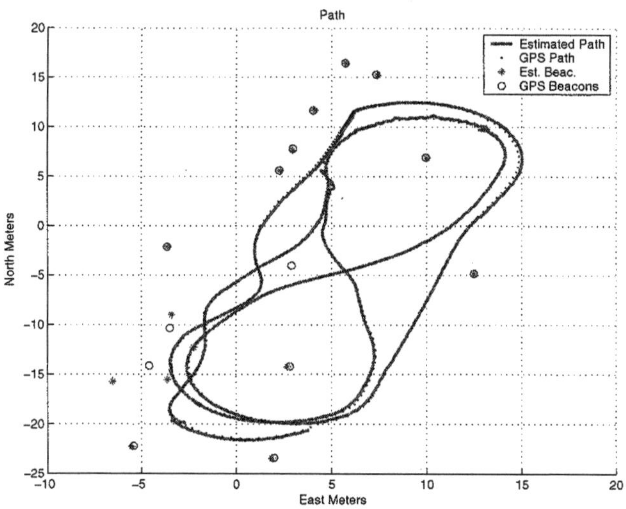

(b) UKF Final map

Fig. 4.3. Comparison of the Extended Kalman Filter and the Unscented Kalman Filter on the Car Park dataset from the University of Sydney. Final vehicle path.

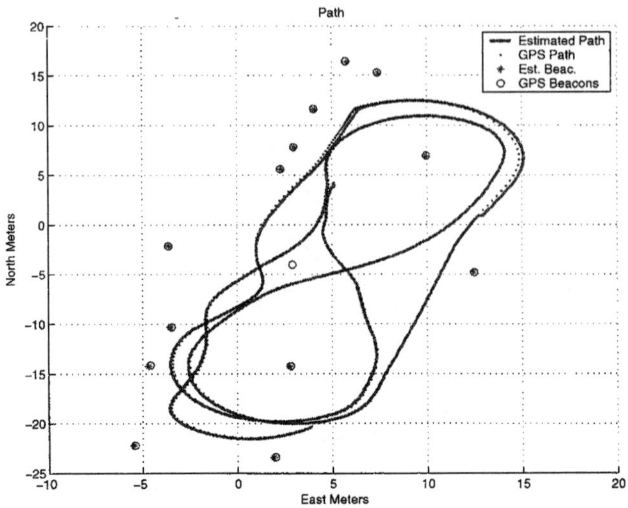

(a) UT Final map

Fig. 4.4. Unscented Transformation of Vehicle States only on the Car Park dataset from the University of Sydney. Final vehicle path.

5

Simultaneous Localization, Control and Mapping

The issue of combining control and estimation together during SLAM has in general been addressed with the idea of online high level trajectory generation. For example, by studying which vehicle maneuvers would most effectively reduce localization uncertainty [25, 78], or what maneuvers would provide the greatest reward in terms of exploration gain [31]; by incorporating visual servoing techniques [20], or by implementing a PD controller over an A* searched trajectory [86].

However, these strategies cannot guarantees in general that the planned trajectories will be followed accurately, in a systems theoretical sense, spite the duality between regulation and linear estimation. That is precisely the focus of this chapter: to provide a unified approach to Multirobot Simultaneous Localization, Control and Mapping, from an estimation-control theoretic perspective, that would generate the necessary low-level control commands to accurately follow a high level planned trajectory, and that would guarantee that both the controller and the estimator are asymptotically stable. Given that observability is a requisite for stable SLAM [11, 10, 59], it is of uttermost importance to guarantee stability of the closed loop system as well. That is, not only during estimation, but also during vehicle control.

More specifically, by using a nonlinear control technique called Feedback Linearization over the EKF state estimates, we are capable of accurately following any multirobot trajectory parameterized in time, while at the same time building an optimally estimated map. Such trajectory could be generated on line, for example, to reduce estimation error, or to maximize exploration gain. Furthermore, extending the Separability principle for the LQG regulator and the Kalman estimator to the feedback linearization scheme, we are able to decouple control error from estimation error, thus guaranteeing stability both for the controller as well as the estimator.

5.1 Linear Quadratic Gaussian Regulation

In this Section we show how to use the optimal vehicle state estimate provided by the Kalman filter to drive the robot to a desired location. In control theory, the problem is known as regulation, and the preferred tool to optimally solve it is the Linear Quadratic Gaussian regulator, for the linear (linearized) case [40].

The idea behind the LQG regulator is to optimally compute an input $\mathbf{u}_k$ to drive the vehicle to a desired location by minimizing a quadratic performance index of the form

$$J = E\left[\sum_{i=0}^{k-1} \mathbf{x}_i^\top \mathbf{Q}_1 \mathbf{x}_i + \mathbf{u}_i^\top \mathbf{Q}_2 \mathbf{u}_i\right] \tag{5.1}$$

where $\mathbf{Q}_1$ and $\mathbf{Q}_2$ are chosen *psd.* The minimization of the performance index (5.1) would drive the state $\mathbf{x}$ to zero, but the extension to a general desired nonzero state is straightforward.

The performance index in (5.1) can be decomposed in two terms, one for minimizing the state estimate and the input, and the second for minimizing the estimation error. Given that the estimation error does not depend on the control input $\mathbf{u}$, it is chosen to minimize the first part of J only. The control law for such regulator is

$$\mathbf{u} = -\mathbf{L}\mathbf{x}_{k|k} \tag{5.2}$$

$$\mathbf{L} = (\mathbf{Q}_2 + \mathbf{G}^\top \mathbf{S}\mathbf{G})^{-1}\mathbf{G}^\top \mathbf{S}\mathbf{F} \tag{5.3}$$

$$\mathbf{S} = \mathbf{Q}_1 + \mathbf{L}^\top \mathbf{Q}_2 \mathbf{L} + (\mathbf{F} - \mathbf{G}\mathbf{L})^\top \mathbf{S}(\mathbf{F} - \mathbf{G}\mathbf{L}) \tag{5.4}$$

The separation of the LQG in two parts, the optimal state estimation, and the optimal controller gives a Kalman filter independent of the matrices $\mathbf{Q}_1$ and $\mathbf{Q}_2$, which specify the optimal controller. In the same way that the optimal control gain $\mathbf{L}$ does not depend on the statistics $\mathbf{P}$, $\mathbf{Q}$, and $\mathbf{R}$ of the random noises.

Furthermore, the separation principle allows writing the closed loop system dynamics as

$$\begin{bmatrix} \mathbf{x}_{k+1} \\ \tilde{\mathbf{x}}_{k+1} \end{bmatrix} = \begin{bmatrix} \mathbf{F} - \mathbf{G}\mathbf{L} & \mathbf{G}\mathbf{L} \\ \mathbf{0} & \mathbf{F} - \mathbf{K}\mathbf{H}\mathbf{F} \end{bmatrix} \begin{bmatrix} \mathbf{x}_k \\ \tilde{\mathbf{x}}_k \end{bmatrix} + \begin{bmatrix} \mathbf{G} & \mathbf{0} \\ \mathbf{G} - \mathbf{K}\mathbf{H}\mathbf{G} & -\mathbf{K} \end{bmatrix} \begin{bmatrix} \mathbf{v}_k \\ \mathbf{w}_k \end{bmatrix} \tag{5.5}$$

The eigenvalues of the closed-loop system are given by those of the state-feedback regulator dynamic $\mathbf{F}-\mathbf{G}\mathbf{L}$ together with those of the state-estimator dynamics $\mathbf{F} - \mathbf{K}\mathbf{H}\mathbf{F}$. In case these both matrices are stable, then so is the closed-loop. For a fully observable monobot, it is straightforward to verify $\mathbf{F} - \mathbf{K}\mathbf{H}\mathbf{F}$ always stable, and the vehicle part of $\mathbf{F} - \mathbf{G}\mathbf{L}$ stable for any positive $\mathbf{L}$.

In order to show the feasibility of using LQG regulation during SLAM, we simulate first the case of the monobot. Figures 5.2 - 5.4 show the results of

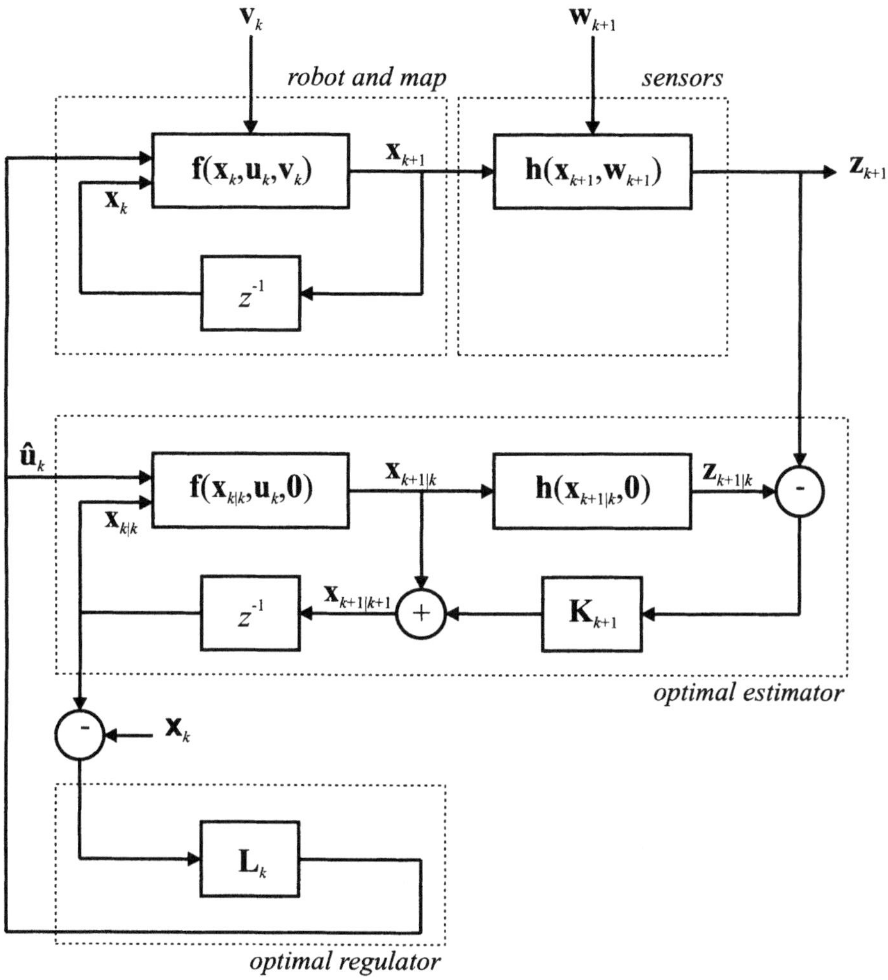

Fig. 5.1. Structure of the optimal state estimator and regulator in SLAM.

applying optimal control to the fully observable SLAM model with artificial process noise from Figures 2.17 and 2.18.

Now, we simulate an environment with 16 landmarks over a $600m^2$ area for the all terrain planar vehicle from Figure 1.7. Section 1.2 includes closed form expressions for the actual vehicle and measurement models used. Note that the pose of the robot, and hence, the control point is located apart from the vehicle axis of rotation in order to avoid singularities in the computation of the Jacobians. The vehicle state and control point is chosen at the origin of a laser range scanner placed on the front of the vehicle, thus simplifying the measurement model.

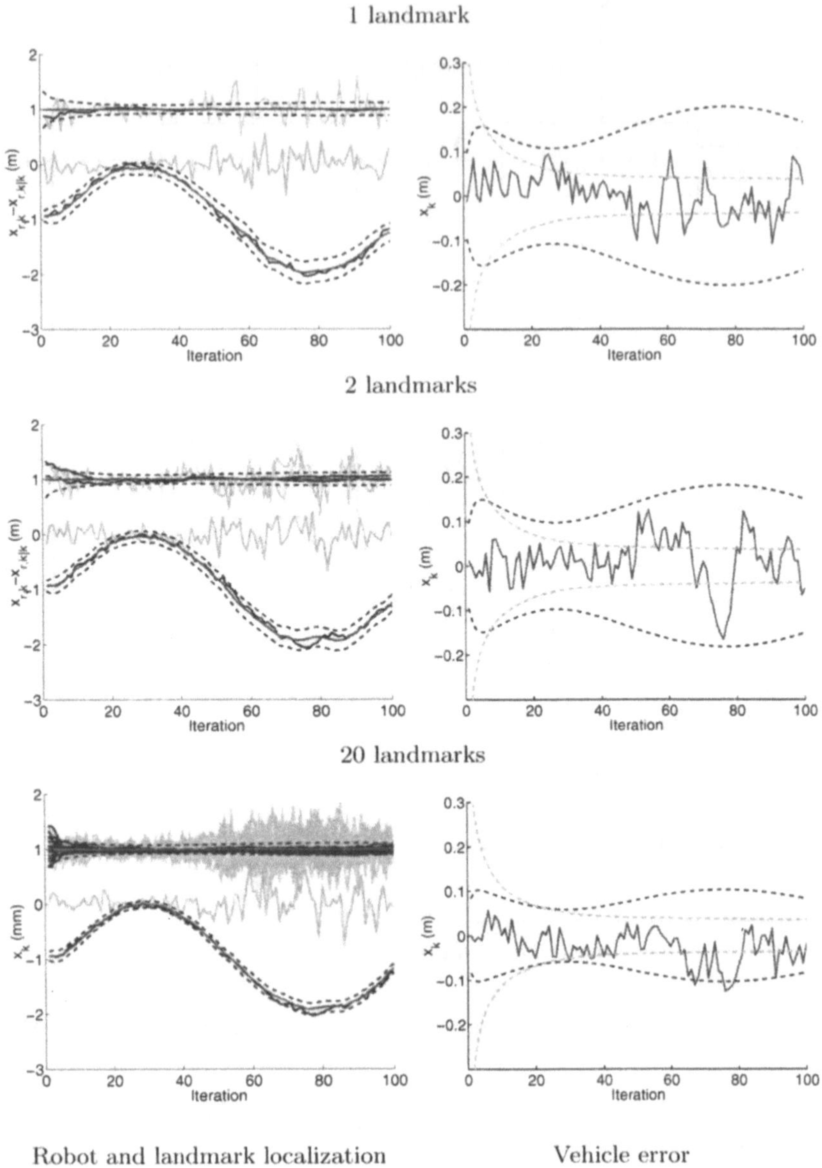

Fig. 5.2. Optimal state estimation and regulation for a monobot in a sinusoidal path starting from $\mathbf{x}_{r,0|0} = -1m$, and with 100 iterations.

In the plots shown in Figure 5.5 the vehicle is driven with the LQG regulator from an initial location at (-7,0) meters to the point (10,10) meters, while at the same time building a map of the environment, and using the

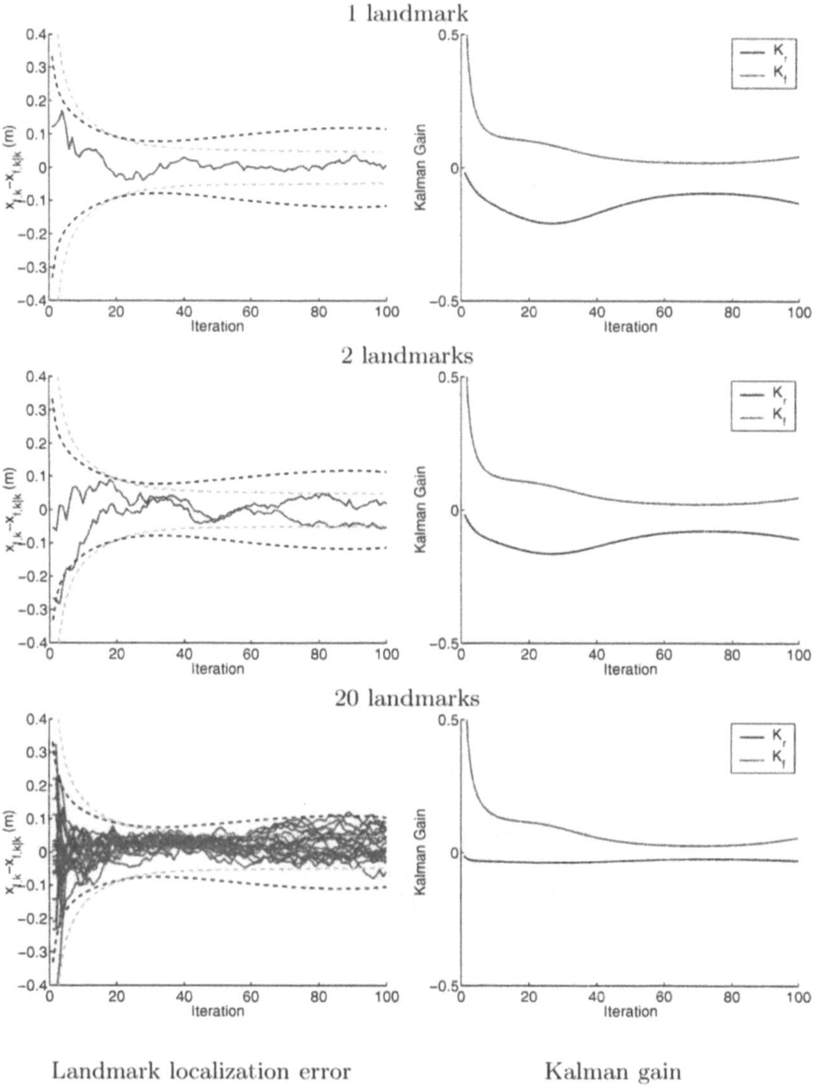

Fig. 5.3. Optimal state estimation and regulation for a monobot in a sinusoidal path starting from $\mathbf{x}_{r,0|0} = -1m$, and with 100 iterations.

revised Kalman estimates to recompute the optimal control gain. Figure 5.6 shows plots of the vehicle state estimate, the estimation error, and the input command. Moreover, Figure 5.8 shows a runtime plot of the eigenvalues of $\mathbf{F} - \mathbf{GL}$ evaluated with the plant Jacobians $\mathbf{F}$ and $\mathbf{G}$. The plot shows the control strategy to be not only optimal by duality from the EKF, but also stable, even when linearizations are used.

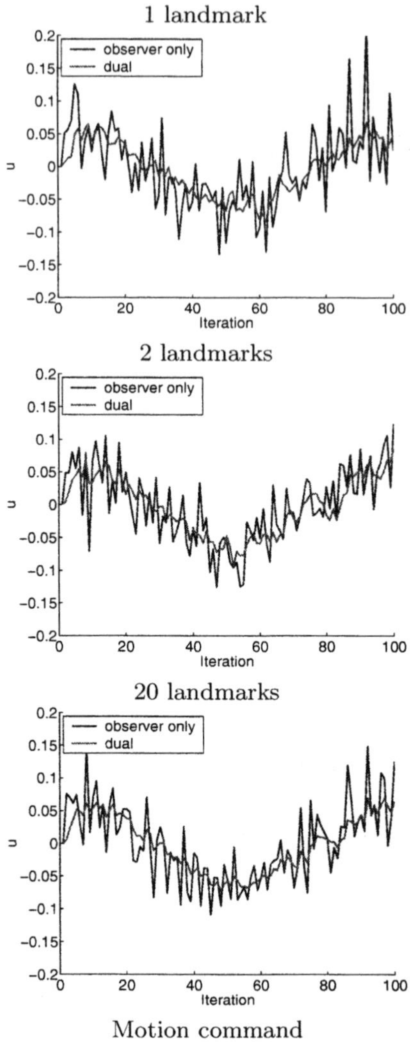

Fig. 5.4. Optimal state estimation and regulation for a monobot in a sinusoidal path starting from $\mathbf{x}_{r,0|0} = -1m$, and with 100 iterations.

5.2 The EKF for Multirobot SLAM

To make our control problem more exciting, in the coming discussion, we will set up the SLAM and control equations not just for one, but for a troupe of robots. Following the notation at the beginning of this book, the motion of the entire troupe of robots and the map measurements are governed by the discrete time stochastic state transition model

$$\mathbf{x}_{k+1} = \mathbf{f}(\mathbf{x}_k, \mathbf{u}_k, \mathbf{v}_k) \tag{5.6}$$

$$\mathbf{z}_k = \mathbf{h}(\mathbf{x}_k) + \mathbf{w}_k \tag{5.7}$$

The state $\mathbf{x}_k = [\mathbf{x}_{r_1,k}^\top, \ldots, \mathbf{x}_{r_t,k}^\top, \mathbf{x}_{f(1)}^\top, \ldots, \mathbf{x}_{f(n)}^\top]^\top$ contains the pose of the robots $\mathbf{x}_{r_1}, \ldots, \mathbf{x}_{r_t}$ at time step k, and a vector of stationary map features $\mathbf{x}_{f(1)}, \ldots, \mathbf{x}_{f(n)}$. The input vector $\mathbf{u}_k = [\mathbf{u}_{r_1,k}^\top, \ldots, \mathbf{u}_{r_t,k}^\top]^\top$ is a multi-vehicle control command, $\mathbf{v}_k = [\mathbf{v}_{r_1,k}^\top, \ldots, \mathbf{v}_{r_t,k}^\top]^\top$ is the plant noise, $\mathbf{w}_k = [\mathbf{w}_{f(1),k}^\top, \ldots, \mathbf{w}_{f(n),k}^\top]^\top$ is the sensor noise, and both are Gaussian random vectors with zero mean and block diagonal covariance matrices $\mathbf{Q}$ and $\mathbf{R}$, respectively. In the same manner as with the one robot case, an optimal estimate of (5.6), in a least squares sense, is given by the expression

$$\mathbf{x}_{k+1|k+1} = \mathbf{f}(\mathbf{x}_{k|k}, \mathbf{u}_k, \mathbf{0}) + \mathbf{K}(\mathbf{z}_{k+1} - \mathbf{h}(\mathbf{f}(\mathbf{x}_{k|k}, \mathbf{u}_k, \mathbf{0}))) \tag{5.8}$$

with covariance

$$\mathbf{P}_{k+1|k+1} = \mathbf{F}\mathbf{P}_{k|k}\mathbf{F}^\top + \mathbf{G}\mathbf{Q}\mathbf{G}^\top - \mathbf{K}[\mathbf{H}[\mathbf{F}\mathbf{P}_{k|k}\mathbf{F}^\top + \mathbf{G}\mathbf{Q}\mathbf{G}^\top]\mathbf{H}^\top + \mathbf{R}]\mathbf{K}^\top \tag{5.9}$$

The Jacobians $\mathbf{F}$ and $\mathbf{G}$ represent first order linearizations of the multi-vehicle model with respect to the state and the plant noise. Similarly, the Jacobian $\mathbf{H}$ contains first order linearizations of the measurement model with respect to the entire state.

5.3 Feedback Linearization

We will now design a controller using feedback linearization for multirobot trajectory tracking. The feedback linearization approach is commonly used to control nonlinear systems by algebraically transforming the system dynamics into a linear one, so that linear control techniques can be applied. It differs from conventional (Jacobian) linearization in that linearization is achieved by exact state transformations and feedback, rather than by linear approximations of the dynamics. The controller presented here however, presents some limitations with respect to nonholonomic kinematic constraints. For a more recent treatment to the control of wheeled vehicles in the presence of estimation uncertainties, see [23].

To apply feedback linearization to the multirobot position part of the state, the system dynamics (5.6) must be described in controllability canonical form. That is, linear with respect to the input $\mathbf{u}(k)$.

$$\mathbf{y}_{k+1} = \mathbf{y}_k + \mathbf{B}(\mathbf{u}_k + \mathbf{v}_k) \tag{5.10}$$

with $\mathbf{y}_k = [x_{r_1,k}, y_{r_1,k}, \ldots, x_{r_t,k}, y_{r_t,k}]^\top$ only the multi-vehicle location part of the state vector.

The nonlinear matrix $\mathbf{B}$ is a function of the multirobot part of the state model (1.54). For our particular case, the first two terms indicate the position of the vehicle, expressed in controllable canonical form, whereas the third term, is the vehicle orientation and is given as an incremental function of the input angular velocity. The nonlinear matrix $\mathbf{B}$ takes the form

$$\mathbf{B}(\theta_k) = \begin{bmatrix} \cos\theta_k & -l\sin\theta_k \\ \sin\theta_k & l\cos\theta_k \end{bmatrix} \tag{5.11}$$

Feedback linearization of the entire multirobot subset of the state vector, that is, including the orientation states, is not possible because in that case, $\mathbf{B}$ would be not invertible, and the resulting pseudo-inverse turns out to be rank-deficient (see [19]).

By choosing a control input of the form

$$\mathbf{u}_k = \mathbf{B}^{-1}(\mathbf{u}'_k - \mathbf{y}_k) \tag{5.12}$$

we can cancel the nonlinearities in that subset of the state, $\mathbf{y}_k$, obtaining a single input-state linear relation

$$\mathbf{y}_{k+1} = \mathbf{u}'_k + \mathbf{B}\mathbf{v}_k \tag{5.13}$$

The term $\mathbf{u}'_k$ in (5.13) is a new input to be determined, that can be chosen using traditional linear control techniques. In this case, we have opted for a control law to track a higher level planned multi-robot trajectory parameterized in time $\mathbf{y}^*_k$, guaranteeing at the same time exponential vehicle location dynamics. That is, by defining the trajectory tracking error as

$$\mathbf{e}_k = \mathbf{y}_k - \mathbf{y}^*_k \tag{5.14}$$

the desired error control dynamics is designed such that

$$\mathbf{e}_{k+1} + \mathbf{Q}_1\mathbf{e}_k = \mathbf{0} \tag{5.15}$$

where $\mathbf{Q}_1$ is constant and positive definite, and as will be seen later, with λ's in $\det(\lambda\mathbf{I} + \mathbf{Q}_1) = 0$ within the unitary circle.

Solving for $\mathbf{y}_{k+1}$ in (5.15), substituting in (5.13), and assuming that the expectations for the estimation error $E[\tilde{\mathbf{y}}_k] = \mathbf{0}$, and the plant noise $E[\mathbf{v}_k] = \mathbf{0}$ hold, we get the control law

$$\mathbf{u}_k = \mathbf{B}^{-1}(\mathbf{y}^*_{k+1} - \mathbf{Q}_1\mathbf{y}^*_k - (\mathbf{Q}_1 + \mathbf{I})\hat{\mathbf{y}}_k) \tag{5.16}$$

The control law $\mathbf{u}_k$ is written as a function of available data. That is, it is a function of the time parameterized multi-vehicle trajectories $\mathbf{y}^*_k$, and of the current multi-vehicle state estimates $\hat{\mathbf{y}}_k$.

Notice that in order to have zero mean estimation error of the vehicle states, SLAM must be fully observable [11]. Filter stability turns out to be a prerequisite for this or any other low level control strategy to be asymptotically stable as well. The intuition is straightforward, to accurately control a troupe of vehicles through a predefined trajectory, one must have means to accurately measure their location at all times.

So the control law (5.16), will stabilize the system about the time parameterized trajectory $\mathbf{y}_k^*$.

In order to validate our feedback control scheme, we write the closed loop equations for the multi-vehicle state and multi-vehicle state estimate error, using the fact that $\mathbf{y}_k = \hat{\mathbf{y}}_k + \tilde{\mathbf{y}}_k$.

$$\mathbf{y}_{k+1} = -\mathbf{Q}_1\mathbf{y}_k + (\mathbf{I} + \mathbf{Q}_1)\tilde{\mathbf{y}}_k + \mathbf{B}\mathbf{v}_k + \mathbf{y}_{k+1}^* + \mathbf{Q}_1\mathbf{y}_k^* \tag{5.17}$$

$$\tilde{\mathbf{x}}_{k+1} = (\mathbf{F} - \mathbf{KHF})\tilde{\mathbf{x}}_k + (\mathbf{G} - \mathbf{KHG})\mathbf{v}_k - \mathbf{K}\mathbf{w}_k \tag{5.18}$$

The separation of the problem in two parts, the optimal state estimation, and the controller, gives a Kalman filter independent of the matrix $\mathbf{Q}_1$, which specifies the control strategy. In the same way that the controller does not depend on the statistics $\mathbf{P}$, $\mathbf{Q}$, and $\mathbf{R}$ of the random noises.

The eigenvalues of the closed-loop system are given by those of the linearized-feedback dynamics $-\mathbf{Q}_1$, together with those of the state estimator dynamics $\mathbf{F} - \mathbf{KHF}$. Only when both matrices are stable, so is the closed-loop. We have designed $\mathbf{Q}_1$ for a stable controller, and for a fully observable estimation problem, it is straightforward to verify $\mathbf{F} - \mathbf{KHF}$ always stable [18, 89].

As mentioned before, given the kinematics constraint of the vehicle model used, the entire vehicle pose including orientation cannot be stabilized, and we have decided to let $\mathbf{y}_k = [x_k, y_k]^\top$ be the Cartesian coordinates of the vehicle location only, at the expense of optimally controlling the vehicle orientation. Our experiments show however, that by controlling the vehicle position only, and letting the vehicle orientation be a free variable, after an initial transient interval, the predefined time parameterized trajectories can still be accurately followed with a troupe of vehicles.

In order to show the feasibility of using Feedback Linearization during multi-robot SLAM, we simulated an environment with 16 landmarks over a $600m^2$ area. The vehicle model used in the simulations corresponds to the all-terrain planar vehicle from Figure 1.7, and is given in Section 1.2. Note that the pose of the robot, and hence, the control point is located apart from the vehicle axis of rotation in order to avoid singularities in the computation of the Jacobians. The vehicle state and control point is chosen at the origin of a laser range scanner placed on the front of the vehicle, thus simplifying the measurement model.

Figure 5.5 presents an simulation for a pair of robots simultaneously tracking two time parameterized circular paths, while performing SLAM. The objective is to track the desired path as accurately as possible. The desired trajectory should come from a higher-level planning strategy. But since that is not the scope of this paper, but to guarantee concurrent tracking and estimation stabilities, simple circular paths are chosen instead.

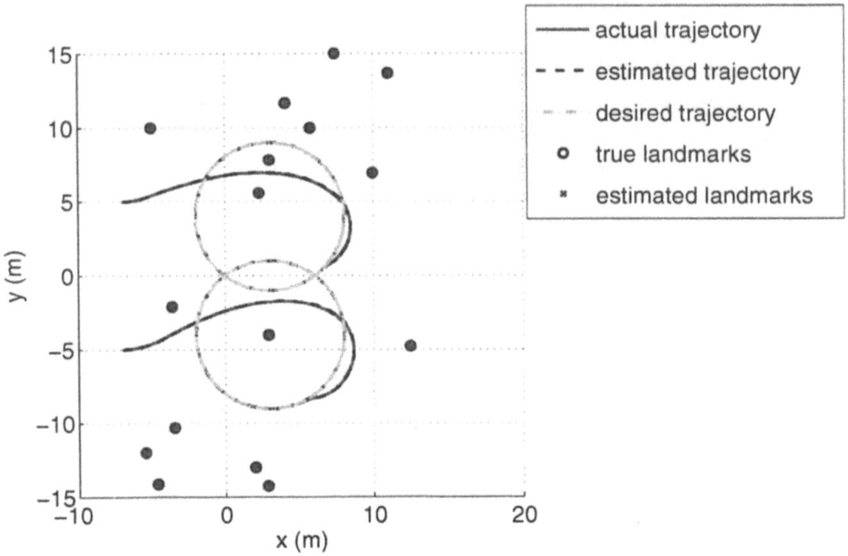

Fig. 5.5. Simultaneous Multirobot Localization, Control, and Mapping

Figure 5.6 shows plots for the vehicle state estimates, the state estimation error, and the history of control commands. Note in the last plot, that when the motion is initiated, the control law chooses a saturated translational velocity to reach the circular path, stabilizing then around 0.6 m/sec. Disregarding the uncontrollability of the angular orientation produces a drastic fluctuation of the angular velocity signal during this initial transient interval, then stabilizing to the desired angular velocity, set at 5 deg/sec.

Finally, Figure 5.8 shows the asymptotic landmark state estimate trace covariances. The plot will look familiar to any experienced SLAM researcher, and specifically shows the decrease in all landmark localization uncertainties as the algorithm proceeds, showing asymptotic convergence of the estimation part of the problem.

5.4 Conclusions

Given the separability between optimal state estimation and regulation, we have been able to present a multi-vehicle low-level control strategy that does not affect the estimation performance of a fully observable EKF based multirobot SLAM: a feedback linearization control strategy that is guaranteed asymptotically stable for close tracking of any time-parameterized high-level computed trajectory. The feasibility of using the approach was validated with simulation results. In order to avoid the initial transient performance of the

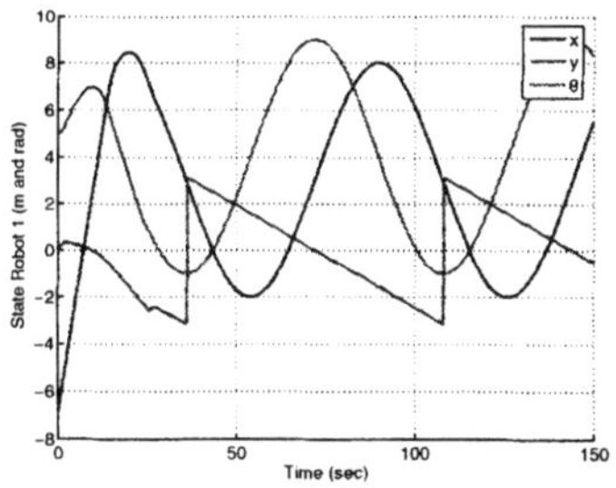

(a) Robot 1 state estimates

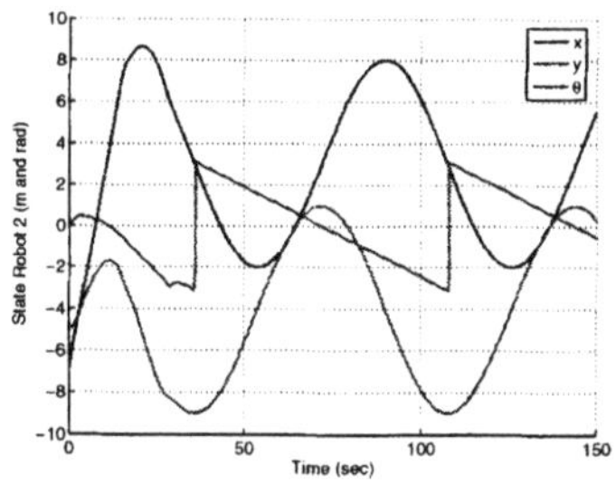

(b) Robot 2 state estimates

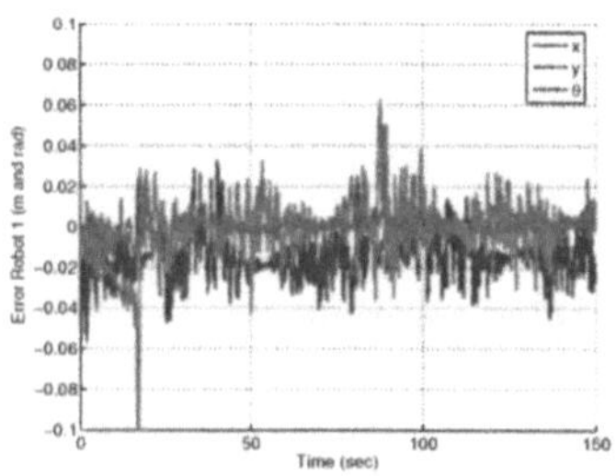

(c) Robot 1 estimation error

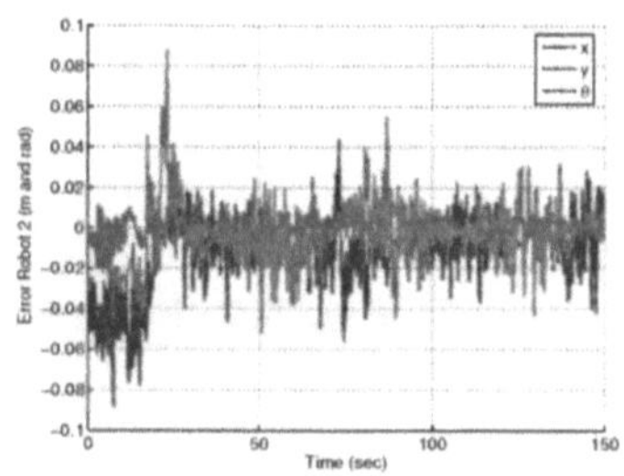

(d) Robot 2 estimation error

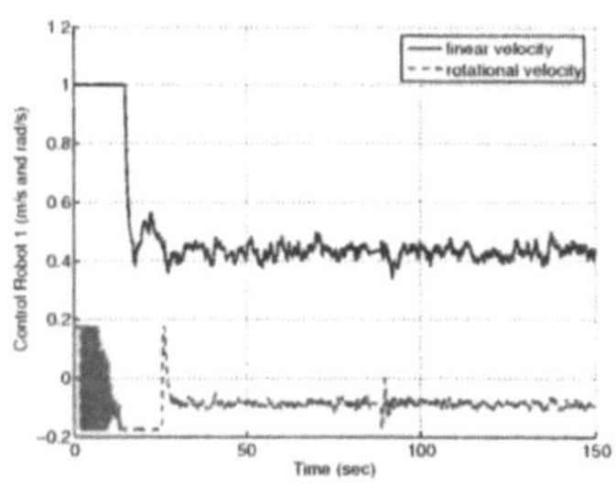

(e) Robot 1 control command

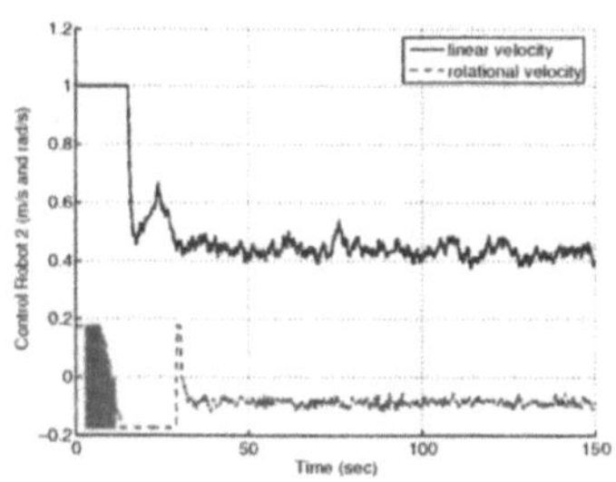

(f) Robot 2 control command

Fig. 5.6. State Estimation and Control using Feedback Linearization.

forward linearization control strategy, the effects of the kinematics constraints on the chosen vehicle model will be further investigated.

The beauty of this chapter is precisely in that it points out the dependence on fully observable SLAM in order to be able to use the SLAM estimates

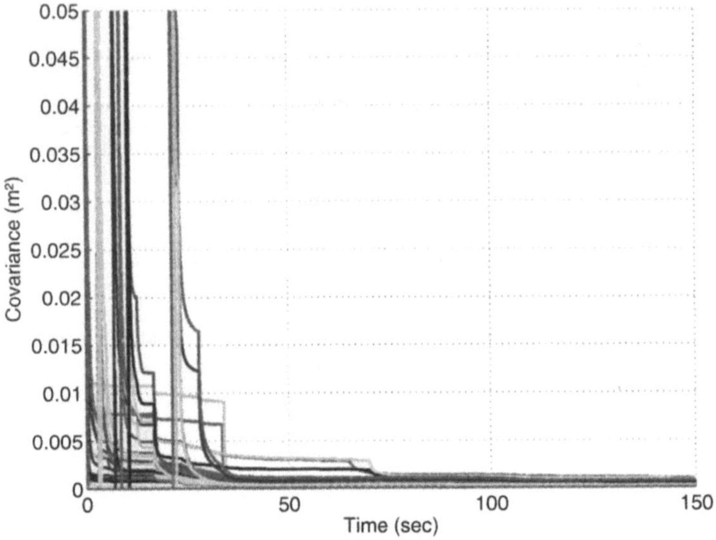

Fig. 5.7. Landmark trace covariances.

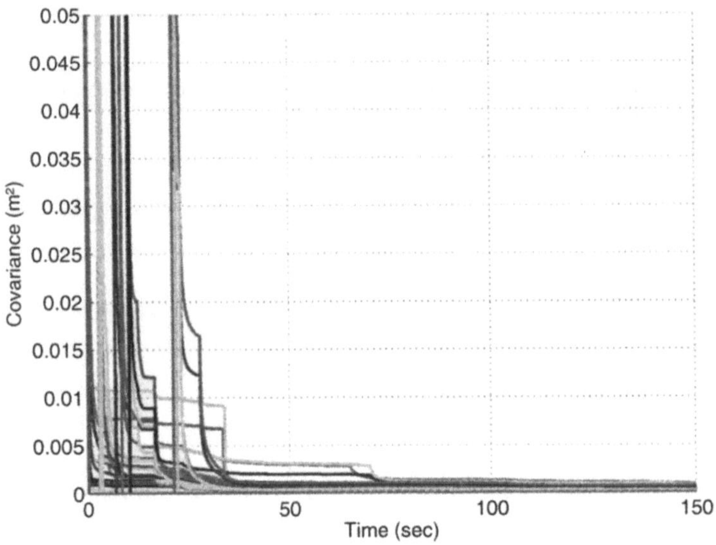

Fig. 5.8. Landmark trace covariances.

as input to any type of controller. Then, both estimation and control can be decoupled and standard techniques such as the ones used here, Kalman filtering for estimation, and feedback linearization for control, are plausible for closing the perception-action-loop in multirobot SLAM.

A

The Kalman Filter

The Kalman Filter developed in the early sixties by R.E. Kalman [57, 58] is a recursive state estimator for partially observed non-stationary stochastic processes. It gives an optimal estimate in the least squares sense of the actual value of a state vector from noisy observations.

Recursive State Estimation

Consider a discrete-time stochastic process

$$\mathbf{x}_{k+1} = \mathbf{f}(\mathbf{x}_k, \mathbf{u}_k, \mathbf{v}_k) \tag{A.1}$$

with system input $\mathbf{u}$ and unmodeled process dynamics plus noise $\mathbf{v}$. The task at hand is to find an estimate of the state vector $\mathbf{x}$. However, $\mathbf{x}$ is only accessible from noise distorted sensor measurements

$$\mathbf{z}_k = \mathbf{h}(\mathbf{x}_k, \mathbf{w}_k) \tag{A.2}$$

in which as with the process model, $\mathbf{w}$ represents observation model inaccuracies and sensor noise.

Recursive state estimation consists on iteratively reconstructing the state vector from our knowledge of the process dynamics, the measurement model, and the sensed data.

Let $\mathbf{x}_{i|j}$, $i \geq j$, be the estimate of the state $\mathbf{x}_i$ using the observation information up to and including time j, $\mathbf{Z}^j = \{\mathbf{z}_0, \ldots, \mathbf{z}_j\}$. Given an estimate $\mathbf{x}_{k|k}$, and the input to the system $\mathbf{u}_k$, the predicted state $\mathbf{x}_{k+1|k}$ is ideally given by the expectation

$$\mathbf{x}_{k+1|k} = E[\mathbf{f}(\mathbf{x}_k, \mathbf{u}_k, \mathbf{v}_k)|\mathbf{Z}^k]. \tag{A.3}$$

We call $\mathbf{x}_{k+1|k}$ the *a priori* estimate of $\mathbf{x}_{k+1}$, and compute it from a noise-free version of Eq. A.1, the estimate $\mathbf{x}_{k|k}$, and the input that hypothetically would drive the process from $\mathbf{x}_k$ to $\mathbf{x}_{k+1}$

$$\mathbf{x}_{k+1|k} = \mathbf{f}(\mathbf{x}_{k|k}, \mathbf{u}_k, \mathbf{0}). \tag{A.4}$$

Combining this result with the discrete-time measurement model from Eq. A.2, we can also predict a noise-free a priori estimate of the sensor measurements

$$\mathbf{z}_{k+1|k} = \mathbf{h}(\mathbf{x}_{k+1|k}, \mathbf{0}). \tag{A.5}$$

By comparing the actual measurement vector $\mathbf{z}_{k+1}$ with the predicted data $\mathbf{z}_{k+1|k}$, we obtain an observation prediction error which in turn is added in a correction term to the a priori state estimate to produce an *a posteriori* state estimate.

$$\mathbf{x}_{k+1|k+1} = \mathbf{x}_{k+1|k} + \mathbf{K}_{k+1}(\mathbf{z}_{k+1} - \mathbf{z}_{k+1|k}). \tag{A.6}$$

The choice of the gain matrix $\mathbf{K}$ usually meets some optimality criteria. In the case of the Kalman Filter, the stochastic nature of the process and measurement dynamics is taken into account in the derivation of $\mathbf{K}$, producing an optimal linear estimator that minimizes the squared error on the expected value of the state estimate $\mathbf{x}_{k+1|k+1}$.

Linear Kalman Filter

Consider the case in which the process and measurement models correspond to a possibly non-stationary[1] discrete-time linear system, and that both the process and sensor noises are zero-mean white[2] and Gaussian with covariance matrices $\mathbf{Q}_k$ and $\mathbf{R}_k$ respectively, then Eqs. A.1 and A.2 become

$$\mathbf{x}_{k+1} = \mathbf{F}_k\mathbf{x}_k + \mathbf{u}_k + \mathbf{v}_k \tag{A.7}$$
$$\mathbf{z}_k = \mathbf{H}_k\mathbf{x}_k + \mathbf{w}_k \tag{A.8}$$

where

$$E[\mathbf{v}_k] = \mathbf{0}, \quad E[\mathbf{v}_k\mathbf{v}_k^\top] = \mathbf{Q}_k, \quad \text{and} \quad E[\mathbf{v}_i\mathbf{v}_j^\top] = \mathbf{0}, \quad \forall i \neq j \tag{A.9}$$
$$E[\mathbf{w}_k] = \mathbf{0}, \quad E[\mathbf{w}_k\mathbf{w}_k^\top] = \mathbf{R}_k, \quad \text{and} \quad E[\mathbf{w}_i\mathbf{w}_j^\top] = \mathbf{0}, \quad \forall i \neq j \tag{A.10}$$

The a priori and a posteriori state estimation errors can be written as

$$\mathbf{e}_{k+1|k} = \mathbf{x}_{k+1} - \mathbf{x}_{k+1|k} \tag{A.11}$$
$$\mathbf{e}_{k+1|k+1} = \mathbf{x}_{k+1} - \mathbf{x}_{k+1|k+1} \tag{A.12}$$

and from the linear model in Eq. A.7, the noise-free a priori state estimate in Eq. A.4 takes the form

$$\mathbf{x}_{k+1|k} = \mathbf{F}_k\mathbf{x}_{k|k} + \mathbf{u}_k. \tag{A.13}$$

---

[1] Hence Kalman filter's beauty, compared to its predecessor the Weiner filter that only works for stationary linear systems.

[2] Temporally uncorrelated and with equal power at all frequencies.

It follows that the a priori state estimate error is given by

$$\mathbf{e}_{k+1|k} = \mathbf{F}_k \mathbf{e}_{k|k} + \mathbf{v}_k. \tag{A.14}$$

Substituting Eq A.6 and the observation models

$$\mathbf{z}_{k+1|k} = \mathbf{H}_{k+1} \mathbf{x}_{k+1|k}$$

and

$$\mathbf{z}_{k+1} = \mathbf{H}_{k+1} \mathbf{x}_{k+1} + \mathbf{w}_{k+1}$$

in Eq. A.12, we obtain a recursive expression for the a posteriori state estimation error

$$\mathbf{e}_{k+1|k+1} = \mathbf{e}_{k+1|k} - \mathbf{K}_{k+1}(\mathbf{H}_{k+1} \mathbf{e}_{k+1|k} + \mathbf{w}_{k+1}). \tag{A.15}$$

The state error covariances are given by the expectations of the square of the state errors.

$$\mathbf{P}_{k+1|k} = E[\mathbf{e}_{k+1|k} \mathbf{e}_{k+1|k}{}^{\top}] \tag{A.16}$$

$$\mathbf{P}_{k+1|k+1} = E[\mathbf{e}_{k+1|k+1} \mathbf{e}_{k+1|k+1}{}^{\top}]. \tag{A.17}$$

Substituting Eq. A.14 in Eq. A.16 and taking the expectations on $\mathbf{v}$, we get the following expression for the a priori state error covariance

$$\mathbf{P}_{k+1|k} = \mathbf{F}_k \mathbf{P}_{k|k} \mathbf{F}_k{}^{\top} + \mathbf{Q}_k. \tag{A.18}$$

For simplicity of notation, in the sequel we rewrite the dependencies $(k+1|k)$ and $(k+1|k+1)$ as $^{\ominus}$ and $^{\oplus}$ respectively, and when no step reference is provided, $(k+1)$ is assumed. Substituting Eq. A.12 in Eq. A.17 and taking the expectations on $\mathbf{w}$ and $\mathbf{e}^{\ominus}$, the a posteriori error covariance takes the form

$$\mathbf{P}^{\oplus} = \mathbf{P}^{\ominus} - \mathbf{P}^{\ominus}\mathbf{H}^{\top}\mathbf{K}^{\top} - \mathbf{K}\mathbf{H}\mathbf{P}^{\ominus} + \mathbf{K}(\mathbf{H}\mathbf{P}^{\ominus}\mathbf{H}^{\top} + \mathbf{R})\mathbf{K}^{\top}. \tag{A.19}$$

The gain matrix $\mathbf{K}$ is chosen to minimize the a posteriori error covariance. Making the derivative of the trace of $\mathbf{P}^{\oplus}$ with respect to $\mathbf{K}$ equal to $\mathbf{0}$, and solving for $\mathbf{K}$ we get the optimal gain for the computation of Eq. A.6, i.e., the Kalman gain

$$\mathbf{K} = \mathbf{P}^{\ominus}\mathbf{H}^{\top}(\mathbf{H}\mathbf{P}^{\ominus}\mathbf{H}^{\top} + \mathbf{R})^{-1}. \tag{A.20}$$

Substituting Eq. A.20 back in Eq. A.19 reduces $\mathbf{P}^{\oplus}$ to the well known form

$$\mathbf{P}^{\oplus} = \mathbf{P}^{\ominus} - \mathbf{K}\mathbf{H}\mathbf{P}^{\ominus}. \tag{A.21}$$

By inspecting the Kalman filter equations the behavior of the filter agrees with our intuition. The Kalman gain is proportional to the uncertainty in the state estimate and inversely proportional to that in the measurements. If sensor readings are very uncertain, and the state estimate is relatively precise, then the Kalman gain has little impact on the update of the state estimate in Eq. A.6, and the system relies heavily on the system model. If, on the other hand, the uncertainty in the measurement is small and that in the state estimate is large, then $\mathbf{K}$ is also large, thus trusting more in sensor measurements for the correction of the state estimate.

However, when sensor measurements are uncertain the second term in Eq. A.21 is small and the state estimate error covariance sees little reduction. Conversely, accurate sensor measurements contribute considerably in reducing the state estimation error.

Given the initial conditions $\mathbf{x}_{0|0}$ and $\mathbf{P}_{0|0}$, the complete recursion in the Kalman filter is computed iteratively with the following steps:

- Predict the a priori state, error covariance, and observation estimates

$$\begin{aligned}\mathbf{x}^{\ominus} &= \mathbf{F}_k\mathbf{x}_{k|k} + \mathbf{u}_k\\ \mathbf{P}^{\ominus} &= \mathbf{F}_k\mathbf{P}_{k|k}\mathbf{F}_k^{\top} + \mathbf{Q}_k\\ \mathbf{z}^{\ominus} &= \mathbf{H}\mathbf{x}^{\ominus}\end{aligned}$$

- Compute the Kalman gain and correct the state and state error covariance estimates

$$\begin{aligned}\mathbf{K} &= \mathbf{P}^{\ominus}\mathbf{H}^{\top}(\mathbf{H}\mathbf{P}^{\ominus}\mathbf{H}^{\top} + \mathbf{R})^{-1}\\ \mathbf{x}^{\oplus} &= \mathbf{x}^{\ominus} + \mathbf{K}(\mathbf{z} - \mathbf{z}^{\ominus})\\ \mathbf{P}^{\oplus} &= \mathbf{P}^{\ominus} - \mathbf{K}\mathbf{H}\mathbf{P}^{\ominus}\end{aligned}$$

Extended Kalman Filter

Consider now the case when the process and observation models in Eqs. A.1 and A.2 are non-linear. The Extended Kalman Filter (EKF) provides a solution by linearizing the process about the current state, and linearizing the measurement model about the predicted observation.

The linearization of $\mathbf{f}$ about the current estimate $\mathbf{x}_{k|k}$ can be formulated as a Taylor series with the higher order terms dropped, that is:

$$\mathbf{x} \approx \mathbf{x}^{\ominus} + \nabla\mathbf{f}_{\mathbf{x}}(\mathbf{x}_k - \mathbf{x}_{k|k}) + \nabla\mathbf{f}_{\mathbf{v}}\mathbf{v}_k.$$

Similarly, the linearization of the observation model takes the form

$$\mathbf{z} \approx \mathbf{z}^{\ominus} + \nabla\mathbf{h}_{\mathbf{x}}(\mathbf{x} - \mathbf{x}^{\ominus}) + \nabla\mathbf{h}_{\mathbf{w}}\mathbf{w}.$$

The noise-free estimates $\mathbf{x}^{\ominus}$ and $\mathbf{z}^{\ominus}$ are given in Eqs. A.4 and A.5, and the various Jacobian matrices contain the partial derivatives of $\mathbf{f}$ and $\mathbf{h}$ with respect to $\mathbf{x}$ and the noises $\mathbf{v}$ and $\mathbf{w}$

$$
\begin{aligned}
\nabla \mathbf{f_x} &= \left.\frac{\partial \mathbf{f}}{\partial \mathbf{x}}\right|_{(\mathbf{x}_{k|k},\mathbf{u}_k,\mathbf{0})} \\
\nabla \mathbf{f_v} &= \left.\frac{\partial \mathbf{f}}{\partial \mathbf{v}}\right|_{(\mathbf{x}_{k|k},\mathbf{u}_k,\mathbf{0})} \\
\nabla \mathbf{h_x} &= \left.\frac{\partial \mathbf{h}}{\partial \mathbf{x}}\right|_{(\mathbf{x}^{\ominus},\mathbf{0})} \\
\nabla \mathbf{h_w} &= \left.\frac{\partial \mathbf{h}}{\partial \mathbf{w}}\right|_{(\mathbf{x}^{\ominus},\mathbf{0})}
\end{aligned}
$$

Following the same discussion as in the previous section but with this new linear model, it is easy to show how the complete recursion for the Extended Kalman Filter involves the following steps:

- Predict the a priori state and observation estimates as well as the a priori state error covariance estimate

$$
\begin{aligned}
\mathbf{x}^{\ominus} &= \mathbf{f}(\mathbf{x}_{k|k}, \mathbf{u}_k, \mathbf{0}) \\
\mathbf{P}^{\ominus} &= \nabla \mathbf{f_x} \mathbf{P}_{k|k} \nabla \mathbf{f_x}^{\top} + \nabla \mathbf{f_v} \mathbf{Q}_k \nabla \mathbf{f_v}^{\top} \\
\mathbf{z}^{\ominus} &= \mathbf{h}(\mathbf{x}^{\ominus}, \mathbf{0})
\end{aligned}
$$

- Compute the Kalman gain and correct the state and state error covariance estimates

$$
\begin{aligned}
\mathbf{K} &= \mathbf{P}^{\ominus} \nabla \mathbf{h_x}^{\top} (\nabla \mathbf{h_x} \mathbf{P}^{\ominus} \nabla \mathbf{h_x}^{\top} + \nabla \mathbf{h_w} \mathbf{R} \nabla \mathbf{h_w}^{\top})^{-1} \\
\mathbf{x}^{\oplus} &= \mathbf{x}^{\ominus} + \mathbf{K}(\mathbf{z} - \mathbf{z}^{\ominus}) \\
\mathbf{P}^{\oplus} &= \mathbf{P}^{\ominus} - \mathbf{K} \nabla \mathbf{h_x} \mathbf{P}^{\ominus}
\end{aligned}
$$

It is important to note however, that the linearization of the nonlinear process and measurement models in the EKF does not preserve the distributions of the state and measurement random variables as normal. This may lead to difficulties in the implementation and tuning of the EKF, making it only reliable for systems that are almost linear on the time scale interval $(k, k+1)$.

Conditioning

It turns out that the recursion in Eq. A.21 is ill-conditioned. As the filter converges, the cancelling of significant digits on $\mathbf{P}^{\oplus}$ may lead to asymmetries or to a non *positive semi definite* (psd) matrix, which cannot be true from the definition in Eq. A.17 of the a posteriori error covariance matrix.

An algebraic manipulation that guarantees $\mathbf{P}^{\oplus}$ psd is obtained by multiplying Eq. A.20 by $(\mathbf{H}\mathbf{P}^{\ominus}\mathbf{H}^{\top} + \mathbf{R})\mathbf{K}^{\top}$, rearranging terms

$$
\mathbf{K}\mathbf{H}\mathbf{P}^{\ominus}\mathbf{H}^{\top}\mathbf{K}^{\top} - \mathbf{P}^{\ominus}\mathbf{H}^{\top}\mathbf{K}^{\top} + \mathbf{K}\mathbf{R}\mathbf{K}^{\top} = \mathbf{0} \tag{A.22}
$$

and adding Eq. A.22 into Eq. A.21

$$\mathbf{P}^{\oplus} = (\mathbf{I} - \mathbf{K}\mathbf{H})\mathbf{P}^{\ominus}(\mathbf{I} - \mathbf{K}\mathbf{H})^{\top} + \mathbf{K}\mathbf{R}\mathbf{K}^{\top} \tag{A.23}$$

The recursivity in Eq. A.23 is known as the Joseph form of the a posteriori error covariance matrix, and given its quadratic nature it is obviously psd.

Sequential Innovation

When combining information from multiple sensors or from multiple data sources, the observation vector $\mathbf{z}$ can be seen as a collection of n independent measurements $\mathbf{z}^{(i)}$ coming from the same number of independent sources at any particular time instance $(k+1)$.

It is possible to process each of these observations independently provided $\mathbf{R}$ is block diagonal. This is, when the set of measurements taken at the same time interval are uncorrelated. Even when the measurements are correlated, they may always be transformed into uncorrelated data which then may be treated sequentially. The process is called *whitening* (see [39]).

Starting from $\mathbf{x}^{\oplus,0} = \mathbf{x}^{\ominus}$, and $\mathbf{P}^{\oplus,0} = \mathbf{P}^{\ominus}$, the a posteriori state estimate is iteratively given by

$$\mathbf{x}^{\oplus,i} = \mathbf{x}^{\oplus,i-1} + \mathbf{K}_{(i)}\left(\mathbf{z}^{(i)} - \mathbf{H}^{(i)}\mathbf{x}^{\oplus,i-1}\right).$$

The key advantage of the *sequential innovation* method is that the complexity in the computation of the Kalman gain is reduced considerably. From Eq. A.20

$$\mathbf{K}_{(i)} = \mathbf{P}^{\oplus,i-1}\mathbf{H}^{(i)\top}\left(\mathbf{H}^{(i)}\mathbf{P}^{\oplus,i-1}\mathbf{H}^{(i)\top} + \mathbf{R}^{(i)}\right)^{-1}. \tag{A.24}$$

The required inverse in Eq. A.24 has the dimension of each of the observed variables, and is considerably much smaller than the dimension of the entire measurement vector $\mathbf{z}$ as required in Eq. A.20. When a sensor returns scalar values for each independent measurement, then the inverse in Eq. A.24 becomes just a scalar division.

Given the initial conditions $\mathbf{x}_{0|0}$ and $\mathbf{P}_{0|0}$, the complete sequential innovation Kalman filter recursion is computed with the following steps:

- Predict the a priori state and state error covariance

$$\mathbf{x}^{\ominus} = \mathbf{F}_k\mathbf{x}_{k|k} + \mathbf{u}_k$$
$$\mathbf{P}^{\ominus} = \mathbf{F}_k\mathbf{P}_{k|k}\mathbf{F}_k^{\top} + \mathbf{Q}_k$$

- For each measurement, iteratively compute the corresponding innovation and Kalman gain column and correct the state and state error covariance estimates

$$
\begin{array}{lll}
\text{initialize} & \mathbf{x}^{\oplus,0} & = \mathbf{x}^{\ominus} \\
 & \mathbf{P}^{\oplus,0} & = \mathbf{P}^{\ominus} \\
\forall i & \mathbf{K}^{(i)} & = \mathbf{P}^{\oplus,i-1}\mathbf{H}^{(i)\top}\left(\mathbf{H}^{(i)}\mathbf{P}^{\oplus,i-1}\mathbf{H}^{(i)\top} + \mathbf{R}^{(i)}\right)^{-1} \\
 & \mathbf{x}^{\oplus,i} & = \mathbf{x}^{\oplus,i-1} + \mathbf{K}_{(i)}\left(\mathbf{z}^{(i)} - \mathbf{H}^{(i)}\mathbf{x}^{\oplus,i-1}\right) \\
 & \mathbf{P}^{\oplus,i} & = \mathbf{P}^{\oplus,i-1} - \mathbf{K}_{(i)}\mathbf{H}^{(i)}\mathbf{P}^{\oplus,i-1} \\
\text{restore} & \mathbf{x}^{\oplus} & = \mathbf{x}^{\oplus,n} \\
 & \mathbf{P}^{\oplus} & = \mathbf{P}^{\oplus,n}
\end{array}
$$

Bibliographical Notes

The reader can find thorough discussions on the Kalman Filter in [18, 29, 57, 58, 68, 71, 73, 92], and on its predecessor the Weiner Filter in [42]. One approach to reduce the effect of nonlinearities is to apply iteratively the filter (IEKF) as in [94]. Another solution is to use the Unscented Kalman Filter (UKF), an extension to the EKF that takes into account the nonlinear transformation of means and covariances [48, 55]. Numerical instability may occur even with the Joseph form of the error covariance matrix. An alternative is the use of the square-root Kalman filter (SKF), in which recursive computations for $\mathbf{P}^{\oplus}$ are substituted by equations for a recursion in ${\mathbf{P}^{\oplus}}^{1/2}$ [29]. Sequential innovation in Kalman filtering is discussed in detail in [18, 29, 91, 93].

B

Concepts from Linear Algebra

Properties of positive semi-definite matrices

Each of the following are necessary and sufficient conditions for a real valued symmetric matrix $\mathbf{A}$ to be positive semi-definite *psd*.

- $\mathbf{x}^\top \mathbf{A}\mathbf{x} \geq 0$.
- all the eigenvalues of $\mathbf{A}$ satisfy $\lambda_i \geq 0$.

Moreover, if $\mathbf{A}$ is *psd*

- $\det(\mathbf{A}) \geq 0$.
- any principal submatrix of a *psd* matrix is also *psd*.
- $\mathbf{B}\mathbf{A}\mathbf{B}^\top$ is also *psd*, for any real valued $\mathbf{B}$.
- and if $\mathbf{B}$ is also *psd*, then $(\mathbf{A} + \mathbf{B})$ is also *psd*.
- and if $\mathbf{B}$ is also *psd*, then $\det \mathbf{A} \leq \det(\mathbf{A} + \mathbf{B})$.

Linear subspaces

The four fundamental subspaces spanned by the columns and rows of a real valued matrix $\mathbf{A}$ are

- The column space of $\mathbf{A}$. $\mathrm{Im}\mathbf{A} = \{\mathbf{y} \in \mathbb{R}^m : \exists \mathbf{x} \in \mathbb{R}^n, \mathbf{A}\mathbf{x} = \mathbf{y}\}$.
- The row space of $\mathbf{A}$. $\mathrm{Im}\mathbf{A}^\top = \{\mathbf{y} \in \mathbb{R}^n : \exists \mathbf{x} \in \mathbb{R}^m, \mathbf{A}^\top\mathbf{x} = \mathbf{y}\}$.
- The null space of $\mathbf{A}$. $\mathrm{Ker}\mathbf{A} = \{\mathbf{x} \in \mathbb{R}^n : \mathbf{A}\mathbf{x} = \mathbf{0}\}$.
- The left null space of $\mathbf{A}$. $\mathrm{Ker}\mathbf{A}^\top = \{\mathbf{x} \in \mathbb{R}^m : \mathbf{A}^\top\mathbf{x} = \mathbf{0}\}$.

Inverse of a block matrix

The inverse of the nonsingular block matrix

$$\begin{bmatrix} \mathbf{A}_{11} & \mathbf{A}_{12} \\ \mathbf{A}_{21} & \mathbf{A}_{22} \end{bmatrix}^{-1} = \begin{bmatrix} \mathbf{B}_{11} & \mathbf{B}_{12} \\ \mathbf{B}_{21} & \mathbf{B}_{22} \end{bmatrix}$$

has the partitions

$$\begin{aligned} \mathbf{B}_{11} &= (\mathbf{A}_{11} - \mathbf{A}_{12}\mathbf{A}_{22}^{-1}\mathbf{A}_{21})^{-1} \\ \mathbf{B}_{12} &= -\mathbf{B}_{11}\mathbf{A}_{12}\mathbf{A}_{22}^{-1} \\ \mathbf{B}_{21} &= -\mathbf{A}_{22}^{-1}\mathbf{A}_{21}\mathbf{B}_{11} \\ \mathbf{B}_{22} &= (\mathbf{A}_{22} - \mathbf{A}_{21}\mathbf{A}_{11}^{-1}\mathbf{A}_{12})^{-1} \end{aligned}$$

The matrix inversion lemma

$$(\mathbf{A}^{-1} + \mathbf{B}^\top \mathbf{C}^{-1}\mathbf{B})^{-1} = \mathbf{A} - \mathbf{A}\mathbf{B}^\top(\mathbf{B}\mathbf{A}\mathbf{B}^\top + \mathbf{C})^{-1}\mathbf{B}\mathbf{A}$$

$$(\mathbf{A} + \mathbf{B}\mathbf{C}\mathbf{B}^\top)^{-1} = \mathbf{A}^{-1} - \mathbf{A}^{-1}\mathbf{B}(\mathbf{B}^\top\mathbf{A}^{-1}\mathbf{B} + \mathbf{C}^{-1})^{-1}\mathbf{B}^\top\mathbf{A}^{-1}$$

Matrix inequalities

The matrix inequality

$$\mathbf{A} \geq \mathbf{B}$$

is to be interpreted as follows:

$$\mathbf{C} = \mathbf{A} - \mathbf{B} \geq 0$$

that is, the difference $\mathbf{C}$ of the two matrices is positive semi-definite.

Bibliographical notes

For a comprehensive text on linear algebra the reader is referred to the book by Strang [81].

C

Sigma Points

A set of $p+1$ *sigma* points $\mathcal{S} = \{\mathcal{X}^i, W^i\}$ are deterministically chosen to satisfy a condition set of the form

$$\mathbf{g}(\mathcal{S}, p(\mathbf{x})) = \mathbf{0} \tag{C.1}$$

where $p(\mathbf{x})$ is the *pdf* of $\mathbf{x}$, not necessarily Gaussian, and $\mathbf{g}(\cdot,\cdot)$ determines the information that should be captured about $\mathbf{x}$. For example, in our Gaussian case, to match the mean

$$g_1 = \sum_{i=0}^{p} W^i \mathcal{X}^i - \mathbf{x}_{k|k}, \tag{C.2}$$

the covariance

$$g_2 = \sum_{i=0}^{p} W^i (\mathcal{X}^i - \mathbf{x}_{k|k})(\mathcal{X}^i - \mathbf{x}_{k|k})^\top - \mathbf{P}_{k|k}, \tag{C.3}$$

and skew

$$g_3 = \sum_{i=n}^{p} W^i (\mathcal{X}^i - \mathbf{x}_{k|k})^3. \tag{C.4}$$

One set of points that satisfies such conditions consists in the following symmetrically-distributed set of points [55]:

$$\mathcal{X}^0_{k|k} = \mathbf{x}_{k|k} \tag{C.5}$$

$$W^0 = \frac{\kappa}{n+\kappa} \tag{C.6}$$

$$\mathcal{X}^i_{k|k} = \mathbf{x}_{k|k} + \left(\sqrt{(n+\kappa)\mathbf{P}_{k|k}}\right)_i \tag{C.7}$$

$$\mathcal{X}^{i+n}_{k|k} = \mathbf{x}_{k|k} - \left(\sqrt{(n+\kappa)\mathbf{P}_{k|k}}\right)_i \tag{C.8}$$

$$W^i = W^{i+n} = \frac{1}{2(n+\kappa)} \tag{C.9}$$

where $\left(\sqrt{(n+\kappa)\mathbf{P}_{k|k}}\right)_i$ is the ith row of the Cholesky decomposition $\mathbf{A}$, $n\mathbf{P} = \mathbf{A}^\top \mathbf{A}$, and W^i is the weight associated with the ith point. The term κ is used to scale the third and higher order terms of this set, and n is the augmented state space dimension (states plus noises).

References

1. Proc. 7th int. sym. experimental robotics. In *Proc. 7th Int. Sym. Experimental Robotics*, Honolulu, Dec. 2000.
2. *Proc. 10th Int. Sym. Robot. Res.*, Lorne, Nov. 2001.
3. *Proc. IEEE/RSJ Int. Conf. Intell. Robots Syst.*, Maui, Nov. 2001.
4. *Proc. IEEE Int. Conf. Robot. Automat.*, New Orleans, Apr. 2004.
5. *Proc. IEEE/RSJ Int. Conf. Intell. Robots Syst.*, Sendei, Sep. 2004.
6. J. Andrade-Cetto. *Environment Learning for Indoor Mobile Robots.* PhD thesis, UPC, Barcelona, Apr. 2003.
7. J. Andrade-Cetto and A. Sanfeliu. Concurrent map building and localization on indoor dynamic environments. *Int. J. Pattern Recogn.*, 16(3):361–374, May 2002.
8. J. Andrade-Cetto and A. Sanfeliu. Concurrent map building and localization with landmark validation. In *Proc. 16th IAPR Int. Conf. Pattern Recog.*, volume 2, pages 693–696, Quebec, Aug. 2002. IEEE Comp. Soc.
9. J. Andrade-Cetto and A. Sanfeliu. Temporal landmark validation in CML. In *Proc. IEEE Int. Conf. Robot. Automat.*, pages 1576–1581, Taipei, Sep. 2003.
10. J. Andrade-Cetto and A. Sanfeliu. The effects of partial observability in SLAM. In *Proc. IEEE Int. Conf. Robot. Automat.* [4], pages 397–402.
11. J. Andrade-Cetto and A. Sanfeliu. The effects of partial observability when building fully correlated maps. *IEEE Trans. Robot.*, 21(4):771–777, Aug. 2005.
12. J. Andrade-Cetto, T. Vidal-Calleja, , and A. Sanfeliu. Multirobot C-SLAM: Simultaneous localization, control, and mapping. In *Proc. IEEE ICRA05 Workshop Network Robot Syst.*, Barcelona, Apr. 2005.
13. J. Andrade-Cetto, T. Vidal-Calleja, and A. Sanfeliu. Stochastic state estimation for simultaneous localization and map building in mobile robotics. In V. Kordic, A. Lazinica, and M. Merdan, editors, *Cutting Edge Robotics*, chapter III.3, pages 223–242. Advanced Robotic Systems Press, 2005.
14. J. Andrade-Cetto, T. Vidal-Calleja, and A. Sanfeliu. Unscented transformation of vehicle states in SLAM. In *Proc. IEEE Int. Conf. Robot. Automat.*, pages 324–329, Barcelona, Apr. 2005.
15. G. C. Anousaki and K. J. Kyriakopoulos. Simultaneous localization and map building for mobile robot navigation. *IEEE Robot. Autom. Mag.*, 6(3):42–53, Sep. 1999.

16. N. Ayache and O. Faugeras. Maintaining representations of the environment of a mobile robot. *IEEE Trans. Pattern Anal. Machine Intell.*, 5(6):804–819, Dec. 1989.
17. Y. Bar-Shalom and T. H. Fortmann. *Tracking and Data Association*, volume 179 of *Mathematics in Science and Engineering*. Academic Press, Boston, 1988.
18. Y. Bar-Shalom, X. Rong Li, and T. Kirubarajan. *Estimation with Applications to Tracking and Navigation*. John Wiley & Sons, New York, 2001.
19. R. Brockett. Asymptotic stability and feedback stabilization. In *Differential Geometric Control Theory*, pages 181–191. Birkhauser, 1983.
20. D. Burschka and G. Hager. Principles and practice of real-time visual tracking for navigation and mapping. In *Proc. Int. Workshop on Robot Sensing*, pages 1–8, Graz, May 2004.
21. J. A. Castellanos, J. M. M. Montiel, J. Neira, and J. D. Tardós. The SPMap: A probabilistic framework for simultaneous localization and map building. *IEEE Trans. Robot. Automat.*, 15(5):948–952, Oct. 1999.
22. H. Choset and K. Nagatani. Topological simultaneous localization and mapping (SLAM): Toward exact localization without explicit localization. *IEEE Trans. Robot. Automat.*, 17(2):125–137, Apr. 2001.
23. P. Coelho and U. Nunes. Path-following control of mobile robots in the presence of uncertainties. *IEEE Trans. Robot.*, (2):252–261, Apr. 2005.
24. A. J. Davison and N. Kita. Sequential localisation and map-building for real-time computer vision and robotics. *Robot. Auton. Syst.*, 36:171–183, 2001.
25. A. J. Davison and D. W. Murray. Simultaneous localisation and map-building uisng active vision. *IEEE Trans. Pattern Anal. Machine Intell.*, 24(7):865–880, Jul. 2002.
26. M. Deans and M. Hebert. Experimental comparison of techniques for localization and mapping using a bearing-only sensor. In *Proc. 7th Int. Sym. Experimental Robotics* [1], pages 395–404.
27. R. A. DeCarlo. *Linear Systems. A State Variable Approach with Numerical Implementation*. Prentice Hall, Englewood Cliffs, 1989.
28. K. Demirli and I. B. Türkşen. Mobile robot navigation with generalized modus ponens type fuzzy reasoning. In *Proc. IEEE Int. Conf. Syst., Man, Cybern.*, pages 3724–3729, Vancouver, Oct. 1995.
29. E. D. Dickmanns. Recursive state estimation. Lecture Notes. Spring 1996. Caltech., Apr. 1996.
30. G. Dissanayake, H. Durrant-Whyte, and T. Bailey. A computationally efficient solution to the simultaneous localisation and map building (SLAM) problem. In *Proc. IEEE Int. Conf. Robot. Automat.*, pages 1009–1014, San Francisco, Apr. 2000.
31. M. W. M. G. Dissanayake, P. Newman, S. Clark, H. F. Durrant-Whyte, and M. Csorba. A solution to the simultaneous localization and map building (SLAM) problem. *IEEE Trans. Robot. Automat.*, 17(3):229–241, Jun. 2001.
32. E. R. Dowski. An information theory approach to incoherent information processing. In *Proc. 5th Topical Meeting on Signal Recovery and Synthesis*, OSA Technical Digest, pages 106–108, Mar. 1995.
33. C. Drocourt, L. Delahoche, C. Pegard, and A. Clerentin. Mobile robot localization based on omnidirectional stereoscopic vision perception system. In *Proc. IEEE Int. Conf. Robot. Automat.*, volume 2, pages 1329–1334, Detroit, May 1999.

34. T. Duckett and U. Nehmzow. Mobile robot self-localization and measurement of performance in middle-scale environments. *Robot. Auton. Syst.*, 24(1-2):57–69, Aug. 1998.
35. R. O. Duda, P. E. Hart, and D. G. Stork. *Pattern Classification.* Wiley & Sons, New York, 2nd. edition, 2001.
36. H. Durrant-Whyte, S. Majumder, M. de Battista, and S. Scheding. A Bayesian algorithm for simultaneous localisation and map building. In *Proc. 10th Int. Sym. Robot. Res.* [2].
37. M. Fischler and R. Bolles. Random sample concensus: A paradigm for model fitting with applications to image analysis and automated cartography. *Comm. ACM*, 24:381–385, 1981.
38. D. Fox, W. Burgard, and S. Thrun. Markov localization for mobile robots in dynamic environments. *J. Artif. Intell. Res.*, 30:391–427, Nov. 1999.
39. K. Fukunaga. *Introduction to Statistical Pattern Recognition.* Academic Press, San Diego, 2nd edition, 1990.
40. K. Furuta and A. Sano. *State Variable Methods in Automatic Control.* John Wiley & Sons, Chichester, 1988.
41. P. W. Gibbens, G. M. W. M. Dissanayake, and H. F. Durrant-Whyte. A closed form solution to the single degree of freedom simultaneous localisation and map building (SLAM) problem. In *Proc. IEEE Int. Conf. Decision Control*, pages 408–415, Sydney, Dec. 2000.
42. R. C. Gonzalez and R. E. Woods. *Digital Image Processing.* Addison Wesley, Reading, 1993.
43. J. Guivant and M. Nebot. Navigation in large outdoor unstructured environments. In *Proc. 10th Int. Sym. Robot. Res.* [2].
44. J. E. Guivant and E. M. Nieto. Optimization of simultaneous localization and map-builidng algorithm for real-time implementation. *IEEE Trans. Robot. Automat.*, 17(3):242–257, Jun. 2001.
45. J. E. Guivant and E. M. Nieto. Solving computational and memory requirements of feature-based simultaneous localization and mapping algorithms. *IEEE Trans. Robot. Automat.*, 19(4):749–755, Aug. 2003.
46. R. Hartley and A. Zisserman. *Multiple View Geometry in Computer Vision.* Cambridge University Press, Cambridge, 2000.
47. M. Hashima, F. Hasegawa, S. Kanda, T. Maruyama, and T. Uchiyama. Localization and obstacle detection for a robot for carrying food trays. In *Proc. IEEE/RSJ Int. Conf. Intell. Robots Syst.*, volume 2, pages 345–351, Grenoble, 1997.
48. S. Julier, J. Uhlmann, and H. F. Durrant-Whyte. A new method for the non-linear transformation of means and covariances in filters and estimators. *IEEE Trans. Automat. Contr.*, 45(3):477–482, Mar. 2000.
49. S. J. Julier. A sparse weight Kalman filter approach to simultaneous localisation and map building. In *Proc. IEEE/RSJ Int. Conf. Intell. Robots Syst.* [3], pages 1251–1256.
50. S. J. Julier. The spherical simplex unscented transformation. In *Proc. American Control Conf.*, Denver, Jun. 2003.
51. S. J. Julier. The stability of covariance inflation methods for SLAM. In *Proc. IEEE/RSJ Int. Conf. Intell. Robots Syst.*, pages 2749–2754, Las Vegas, Oct. 2003.

52. S. J. Julier and J. K. Uhlmann. A new extension of the Kalman filter to nonlinear systems. In I. Kadar, editor, *Proc. 11th SPIE Int. Sym. Aerospace/Defense Sensing, Simulation, Controls*, pages 182–193, Orlando, Apr. 1997. SPIE.
53. S. J. Julier and J. K. Uhlmann. A counter example to the theory of simultaneous localization and map building. In *Proc. IEEE Int. Conf. Robot. Automat.*, pages 4238–4243, Seoul, May 2001.
54. S. J. Julier and J. K. Uhlmann. Simultaneous localisation and map building using split covariance intersection. In *Proc. IEEE/RSJ Int. Conf. Intell. Robots Syst.* [3], pages 1257–1262.
55. S. J. Julier and J. K. Uhlmann. Unscented filtering and nonlinear estimation. *Proc. IEEE*, 92(3):401–422, Mar. 2004.
56. T. Kailath. *Linear Systems*. Information and System Sciences Series. Prentice Hall, Englewood Cliffs, 1980.
57. R. E. Kalman. A new approach to linear filtering and prediction problems. *J. Basic Eng.-T. ASME*, pages 35–45, Mar. 1960.
58. R. E. Kalman and R. S. Bucy. New results in linear filtering and prediction theory. *J. Basic Eng.-T. ASME*, pages 95–108, Mar. 1961.
59. J. Kim and S. Sukkarieh. Improving the real-time efficiency of inertial SLAM and understanding its observability. In *Proc. IEEE/RSJ Int. Conf. Intell. Robots Syst.* [5], pages 21–26.
60. A. Kosaka and A. C. Kak. Fast vision-guided mobile robot navigation using model-based reasoning and prediction of uncertainties. *Comput. Vis. Image Und.*, 56(3):271–329, Nov. 1992.
61. Y. D. Kwon and J. S. Lee. A stochastic map building method for mobile robot using 2-d laser range finder. *Auton. Robot.*, 7(2):187–200, 1999.
62. D. Lee and M. Recce. Quantitative evaluation of the exploration strategies of a mobile robot. *Int. J. Robot. Res.*, 16(4):413–447, Aug. 1997.
63. J. J. Leonard, H. F. Durrant-Whyte, and I. J. Cox. Dynamic map building for an autonomous mobile robot. *Int. J. Robot. Res.*, 11(4):286–292, 1992.
64. J. J. Leonard, P. M. Newman, R. J. Rikoski, J. Neira, and J. D. Tardós. Towards robust data association and feature modeling for concurrent mapping and localization. In *Proc. 10th Int. Sym. Robot. Res.* [2].
65. J. J. Leonard and R. Rikoski. Incorporation of delayed decision making into stochastic mapping. In *Proc. 7th Int. Sym. Experimental Robotics* [1].
66. A. Mallet and S. Lacroix. Toward real-time 2d localization in outdoor environments. In *Proc. IEEE Int. Conf. Robot. Automat.*, volume 4, pages 2827–2832, Leuven, May 1998.
67. R. Mandelbaum, G. Kamberova, and M. Mintz. Statistical decision theory for mobile robotics: Theory and application. In *Proc. IEEE Int. Conf. Multisensor Fusion Integration Intell. Syst.*, pages 17–24, Washington, Dec. 1996.
68. P. S. Maybeck. *Stochastic Models, Estimation, and Control*, volume 1. Academic Press, New York, 1979.
69. E. Nebot, J. Guivant, and J. Nieto. ACFR, experimental outdoor dataset, 2002.
70. J. Neira and J. D. Tardós. Data association in stochastic mapping using the joint compatibility test. *IEEE Trans. Robot. Automat.*, 17(6):890–897, Dec. 2001.
71. P. M. Newman. *On the Structure and Solution of the Simultaneous Localisation and Map Building Problem*. PhD thesis, The University of Sydney, Sydney, Mar. 1999.

72. A. Ohya, A. Kosaka, and A. C. Kak. Vision-based navigation by a mobile robot with obstacle avoidance using single-camera vision and ultrasonic sensing. *IEEE Trans. Robot. Automat.*, 14(6):969–978, Dec. 1998.
73. A. Papoulis. *Probability, Random Variables, and Stochastic Processes.* McGraw Hill, New York, 3rd. edition, 1991.
74. I. B. Risteski and K. G. Trenčevski. Principal values and principal subspaces of two subspaces of vector spaces with inner product. *Cont. Algebra Geom.*, 42(1):289–300, 2001.
75. S. Roweis and Z. Ghahramani. A unifying review of linear gaussian models. *Neural Comput.*, 11:305–345, 1999.
76. A. Saffiotti. The uses of fuzzy logic in autonomous robot navigation: a catalogue raisonné. *Soft Computing*, 1(4):180–197, 1997.
77. A. Saffiotti and L. P. Wesley. Perception-based self-localization using fuzzy locations. In *Proc. Int. Workshop Reasoning Uncert. Robot.*, pages 368–385, Amsterdam, 1996.
78. R. Sim, G. Dudek, and N. Roy. Online control policy optimization for minimizing map uncertainty during exploration. In *Proc. IEEE Int. Conf. Robot. Automat.* [4].
79. R. C. Smith and P. Cheeseman. On the representation and estimation of spatial uncertainty. *Int. J. Robot. Res.*, 5(4):56–68, 1986.
80. B. Southall, B. F. Buxton, and J. A. Marchant. Controllability and observability: Tools for kalman filter design. In J. N. Carter and M. S. Nixon, editors, *Proc. British Machine Vision Conf.*, pages 164–173, Southampton, 1998.
81. G Strang. *Linear Algebra and its Applications.* Saunders College Publishing, Fort Worth, 3rd edition, 1988.
82. R. Talluri and J. K. Aggarwal. Mobile robot self-location using model-image feature correspondence. *IEEE Trans. Robot. Automat.*, 12(1):63–77, Feb. 1996.
83. J. D. Tardós, J. Neira, P. M. Newman, and J. J. Leonard. Robust mapping and localization in indoor environments using sonar data. *Int. J. Robot. Res.*, 21(4):311–330, 2002.
84. S. Thrun. Probabilistic algorithms in robotics. *AI Mag.*, 21(4):93–109, 2000.
85. S. Thrun, Y. Liu, D. Koller, A. Y. Ng, Z. Ghahramani, and H. Durrant-Whyte. Simultaneous localization and mapping with sparse extended information filters. *Int. J. Robot. Res.*, 23(7-8):693–716, Jul. 2004.
86. S. Thrun, S. Thayer, W. Whittaker, C. Baker, W. Burgard, D. Ferguson, D. Hannel, M. Montemerlo, A. Morris, Z. Omohundro, and C. Reverte. Autonomous exploration and mapping of abandoned mines. *IEEE Robot. Automat. Mag.*, 11(4):79–91, Dec. 2004.
87. C. Tomasi. Mathematical methods for robotics and vision. Lecture Notes. Spring 2000. Stanford, 2000.
88. T. Vidal-Calleja, J. Andrade-Cetto, and A. Sanfeliu. Conditions for suboptimal filter stability in SLAM. In *Proc. IEEE/RSJ Int. Conf. Intell. Robots Syst.* [5], pages 27–32.
89. T. Vidal-Calleja, J. Andrade-Cetto, and A. Sanfeliu. Estimator stability analysis in SLAM. In *Proc. 5th IFAC/EURON Sym. Intell. Auton. Vehicles*, Lisbon, Jul. 2004.
90. R. Volpe, T. Litwin, and L. Matthies. Mobile robot localization by remote viewing of a colored cylinder. In *Proc. IEEE/RSJ Int. Conf. Intell. Robots Syst.*, volume 1, pages 257–263, Pittsburgh, 1995.

91. G. Welch and G. Bishop. SCAAT: Incremental tracking with inclomplete information. In T. Whitted, editor, *Computer Graphics. Proc. ACM SIGGRAPH Conf.*, pages 333–344, Los Angeles, Aug. 1997. ACM Press.
92. G. Welch and G. Bishop. An introduction to the Kalman filter. Course 8. In *Proc. ACM SIGGRAPH Conf.*, Los Angeles, Aug. 2001. ACM Press.
93. G. F. Welch. *SCAAT: Incremental Tracking with Incomplete Information.* PhD thesis, The University of North Carolina at Chapel Hill, Chapel Hill, Oct. 1996.
94. Z. Zhang. Parameter estimation techniques: A tutorial with application to conic fitting. *Image Vision Comput.*, 15(1):59–76, 1997.

MIX
Papier aus verantwortungsvollen Quellen
Paper from responsible sources
FSC® C105338

If you have any concerns about our products,
you can contact us on
ProductSafety@springernature.com

In case Publisher is established outside the EU,
the EU authorized representative is:
Springer Nature Customer Service Center GmbH
Europaplatz 3, 69115 Heidelberg, Germany

Printed by Libri Plureos GmbH
in Hamburg, Germany